D1631542

THE THINKING DRINKERS ALMANAC

FOR OUR BOYS JOSEPH, SAMUEL, RÈMY AND RORY

An Hachette UK Company
www.hachette.co.uk

First published in Great Britain in 2021 by Kyle Books,
an imprint of Octopus Publishing Group Limited
Carmelite House
50 Victoria Embankment
London EC4Y 0DZ
www.kylebooks.co.uk

ISBN: 978 0 85783 956 5

Publisher: Joanna Copestick
Senior Commissioning Editor: Louise McKeever
Design and illustrations: Paul Palmer-Edwards
Copy-editor: Tara O'Sullivan
Production: Peter Hunt

A Cataloguing in Publication record for this title is available from the British Library

Printed and bound in Great Britain

10 9 8 7 6 5 4 3 2 1

THE

THINKING DRINKERS

ALMANAC

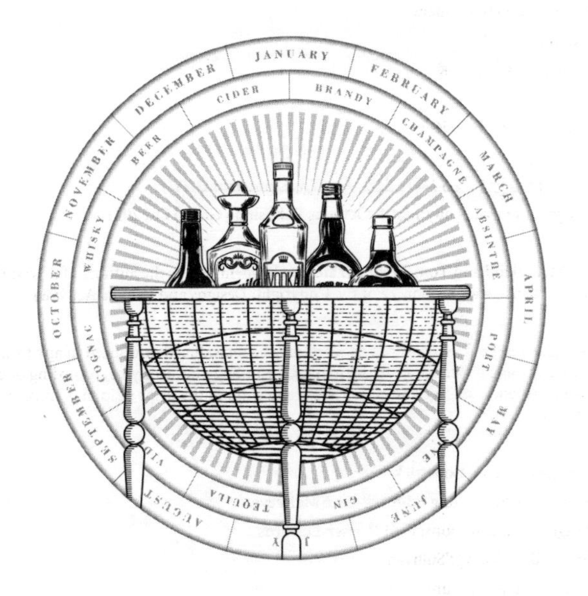

ADVENTUROUS DRINKS AND ECCENTRIC TALES FOR EVERY DAY OF THE YEAR

BEN McFARLAND AND TOM SANDHAM

A note on strength ratings

We are not using official units as they depend on where in the world you are. We also all have a very different tolerance to alcohol. In line with our 'drink less, drink better' mantra, keep to between a total of 15–20 icons a week. We also recommend having alcohol-free days.

1 = *25ml spirit measure (+ mixer)*

2 = *bottle of beer or cider (5% and under ABV) / 150ml (5 fl oz) champagne*

3 = *strong beer or cider (over 5% ABV) / 250ml (8½ fl oz) wine*

4 = *50ml (1½ fl oz) and under spirit measure cocktail*

5 = *over 50ml (1½ fl oz) spirit measure cocktail*

A note on measurements

In order to achieve the right balance of flavours, some of the recipes included use very precise measurements, but feel free to adjust these if you wish, to suit your tastes.

As a general guide:
2.5ml = ½ teaspoon
5ml = 1 teaspoon
15ml = 1 tablespoon

A note on ingredients

Some of the cocktail recipes call for ingredients you may not be familiar with. These can be found online or at specialist suppliers, and are well worth seeking out to take your drinks up a notch or two.

CONTENTS

INTRODUCTION

"Quick, bring me a beaker of wine, so that I may wet my mind and say something clever" – Aristophanes, Ancient Greek Comic.

Yes, that's right. We've introduced this book with a quote from a Classical comedic clever clogs from Ancient Greece.

Pretentious? Bien sur. But, apparently, it gives the book gravitas. Not that the book needs anymore gravitas. If anything, it's got too much gravitas – in fact, on page 11, there's a whole day dedicated to Isaac Newton, the man who actually discovered the stuff – when an apple (which makes cider) bopped him on the head.

It's just one of the 366* entries in this eclectic elbow-bending anthology of discerning drinks and fascinating and unusual facts.

Others include a pair of whisky highballs for 'National Giraffe Day'; a South African sparkling wine served at the inauguration of both Barack Obama and Nelson Mandela; an alpine elixir inspired by a chap who eats aeroplanes; and a gin cocktail created in honour of someone who sewed monkey's testicles onto the face of a Frenchman. If you've always wondered why the Morning Glory Fizz is the ideal toast for the fellow who wrote "The History of Farting" whilst wearing a Fez (him not you), then this is the book you've been looking for.

This weighty tome is a culmination of 40 years dedication to drink – writing, researching and learning about alcohol in all its fabulous, fermented forms. Since meeting while working on a pub newspaper 20 years ago, we have travelled the world visiting distilleries, breweries and winemakers… so you don't have to.

Our unwavering commitment to altruistic imbibing has also taken us to the some of the world's best (and worst) drinking destinations – from swanky cocktail establishments and dirty dive bars to rum shacks, suspect South African shebeens, lovely old pubs and, unfortunately for all involved, faux speakeasies where hipsters take your order using an old typewriter.

Delving deep into drinking history, we've soaked up the people, places and the past that have shaped it – and decanted them into our live comedy tasting shows, broadening the booze horizons of more than 100,000 theatre-goers over the last decade or so.

No-one likes a show off but we genuinely know our stuff – what we could tell you about yeast strains alone could fill a book much bigger than this. But it's not one that anyone would buy. Or read. Or, it turns out, publish.

As experts in alcohol, we're also acutely aware of the potential downfalls of drink and its Jekyll & Hyde character. One minute a soul mate, the next a psychopath, alcohol is a fickle fellow that can flip from faithful friend to fearsome foe in the space of a few sips.

We know that when drink is mishandled and consumed in excess, it can cause all kinds of mischief, misery and mishaps. But when afforded a certain level of reverence and respect, when enjoyed in moderation, we passionately believe that drink is one of life's truly great pleasures.

What's more, it needn't be a 'guilty' one. Booze has had a bad rap over the years, and blamed for a host of society's ills, but we really shouldn't need an excuse to enjoy a glass of something wonderful.

But just in case you do, we've assembled 366 persuasive justifications over the following 272 pages – each accompanied with thrilling tales, absorbing anecdotes and astonishing drinking-related did-you-knows that will impress your friends and make you seem even more interesting and informed than you already are.

Every drink comes with the ingredients and instructions required to make them alongside an indication of alcoholic strength to help you follow the Thinking Drinkers mantra: 'Drink Less, Drink Better'.

We must stress that these alcoholic estimations are simply general guidelines. We don't recommend you drink every day, that would be daft. Instead, use your common sense and please be measured in your measures.

Cheers,
Ben & Tom

* We've thrown in leap year's 29th February for free. You're welcome.

JANUARY

1 JANUARY

BASS TRIANGLE REGISTERS FIRST EVER TRADEMARK (1876)

BASS PALE ALE –

The last thing you want to wake up to with a hangover is the faint whiff of 'Burton snatch' from a pint of Bass Pale Ale. Synonymous with the beers of Burton-on-Trent, the 'snatch' is a subtle, sulphury scent produced by the town's rock-hard water, which is rich in sulphate ions.

When Britain had a swaggering empire, Bass Pale Ale was exported (and imitated) all over the world in brown bottles adorned with an iconic red triangle, which, in 1876, became the UK's first ever trademark.

The 'triangle' famously appears in Édouard Manet's 1882 painting *A Bar at the Folies-Bergère*. The inclusion of a beer brewed in Burton and not Bavaria was indicative of the anti-German sentiment pervading Parisian cafés in the years after the Franco-Prussian War.

Napoleon Bonaparte (see 18 May) thought Bass was '*trés bon*', and tried to set up a similar brewery in Paris – an idea that was abandoned when he realised French water couldn't recreate the inimitable sulphuric 'Burton snatch'.

2 JANUARY

ERIKA ROE STREAKS AT TWICKENHAM (1982)

NAKED LADIES, TWICKENHAM FINE ALES –

Let's face it, rugby can be a bit boring. No one really understands the rules, and for most of the game, the ball disappears under a pile of bodies. Scrums look like a load of fat lads looking for some lost keys, and then it's just half-hearted punch-ups and a small posh man kicking a penalty through some posts.

Back in 1982, sheer boredom inspired Erika Roe to unclasp her bra and run onto the pitch while England were playing Australia. Erika's two-minute topless tour of the Twickenham pitch is regarded as the most iconic streak in history. In memory (or mammary) of this momentous occasion, we're drinking a pint of Naked Ladies from Twickenham Fine Ales, a brewery situated less than a mile from the stadium.

---

3 JANUARY

GEORGE WASHINGTON DUPES THE BRITISH AT THE BATTLE OF PRINCETON (1877)

BOURBON COUNTY STOUT, GOOSE ISLAND –

The Battle of Princeton saw American forces hand the British Army their backsides on a plate.

In this hugely significant battle of the American Revolution, General George Washington and his 5,000 troops hoodwinked a British army led by General Charles Cornwallis, a chubby Old Etonian, by leaving their campfires blazing as they snuck up on the Brits from behind.

Dovetailing Washington's two favourite drinks, porter and whiskey, Goose Island Bourbon County Stout spends nine months hibernating in a variety of wooden casks sourced from an array of American whiskey distilleries. The result is a 15 per cent ABV vortex of vanillin, dark chocolate orange and charred, smoky campfire coffee – the kind that Washington's men would have enjoyed after opening a can of almighty whoop-ass on the Brits.

4 JANUARY

BIRTH OF ISAAC NEWTON (1643)

ASPALL PREMIER CRU –

Sir Isaac Newton was the boffin's boffin. He was regarded by Albert Einstein as the smartest person that ever lived.

Not content with developing the theory of gravity, Newton introduced the laws of motion (which became the foundation for physics) and was also instrumental in the development of telescopes – which, apparently, are well worth looking into.

Famously, while trying to understand optics, he stuck a blunt needle into his own eye – which suggests that he wasn't quite as clever as everyone says he was. An apple falling from a tree *allegedly* inspired his discovery of gravity, so we're enjoying Aspall Premier Cru Cyder, made just down the road from Isaac's birthplace in Suffolk.

It's crisp and beautifully effervescent – and a lot more pleasurable than sticking a blunt needle in your eye.

5 JANUARY

DEATH OF SIR ERNEST SHACKLETON (1922)

MACKINLAY'S RARE OLD HIGHLAND MALT, GLEN MHOR DISTILLERY –

As a young man, Sir Ernest Shackleton used to stand outside pubs and sing about the evils of alcohol.

But in 1907, aged just 33, his attitude to alcohol changed.

Acutely aware that it was a crucial camaraderie catalyst, he embarked on his famous journey to the South Pole – carrying a tonne of booze.

This included 25 cases (300 bottles) of 10-year-old Mackinlay's Rare Old Highland Malt from the Glen Mhor Distillery. In 2006, after a century entombed in ice, three cracked cases were unearthed from beneath Shackleton's Antarctic hut in Cape Royds. The whisky was in fine fettle: its slightly higher ABV of 47.3 per cent had preserved it (at 40 per cent, it would have frozen).

Five years later, Whyte & Mackay released a replica made from a range of Highland malts, including Glen Mhor (which had closed in 1983).

Best served over ice – naturally.

---

6 JANUARY

BIRTH OF JOAN OF ARC (C.1412)

GRAND 75 –

From an early age, Jeanne d'Arc was visited by celestial voices. At first, they mostly talked the kind of tittle-tattle you'd expect a 13-year-old girl to hear: ponies, princesses, spotty boys . . . that kind of thing.

But gradually, these voices got a little more sinister and specific, ordering her to drive the British out of her homeland and bring an end to years of Anglo-French fisticuffs.

As even Greta Thunberg would concede, that's a big ask for a teenage girl, but the voices in Joan's head simply wouldn't shut up about it. Towards the end of her life, while on trial for heresy, she recalled: 'I heard a voice from God . . . about midday, in summer time, in my father's garden . . . from the right side toward the church.'

Instead of asking God what on earth he was doing in her dad's garden, Joan took control of the French army and refused to take any nonsense from her troops. She banned swearing and fraternising with prostitutes, and also famously slapped a Scottish soldier for eating stolen meat (not the last time she had an issue with hot stakes).

While the French hail Joan as a fearless warrior who won the Hundred Years' War, she never actually fought on the battlefield, and was eventually captured by the English, who condemned her to death for wearing men's clothes.

Forced into a dress, clutching a cross and sporting a rather neat 'bob' haircut that remains fashionable to this day, she was burned at the stake aged just 19.

As Joan was born in Champagne, we're celebrating her life with a twist on the 'French 75', using Grand Marnier instead of gin in an iconic champagne cocktail named after a massive artillery field gun used in the First World War.

~

50ml (1½ fl oz) Grand Marnier | Fresh lemon juice | Ice
Champagne, to top up | Orange slice, to garnish

- Shake the Grand Marnier and lemon juice in a cocktail shaker with ice. Strain into a coupe and top up with champagne. Serve, garnished with an orange 'arc' slice.

---

7 JANUARY

GIBSON 'FLYING V' GUITAR DESIGN PATENTED (1958)

GIBSON MARTINI –

Until 1958, all guitars pretty much sported a classic hourglass figure. But then the wonderfully named Seth Lover created the Gibson 'Flying V': a new 'modernistic' pointed design hailed as the most revolutionary electric guitar to have ever been made.

Having failed to fly in the 60s, it enjoyed a hard rock comeback courtesy of Marc Bolan, Billy Gibbons (of ZZ Top), Lenny Kravitz (who famously played one in the video for 'Are You Gonna Go My Way') and Eddie Van Halen (of, er . . . Van Halen).

In the hands of these rock legends, the guitar was a V-shaped gesture to The Man – much like this Martini, which shares the Gibson name and is served with two cocktail onions instead of an olive. An early 20th-century creation, it is said to be named after the buxom Gibson Girls, whose attributes are represented by the two onions.

~

60ml (2 fl oz) gin | 15ml (½ fl oz) dry vermouth
Ice | 2 cocktail onions

- Stir the gin and vermouth together in a mixing glass with some ice, then strain into a Martini glass. Serve garnished with the onions.

---

8 JANUARY

BIRTH OF DAVID BOWIE (1947)

COLDHARBOUR LAGER, BRIXTON BREWERY –

Even though he only got one O Level and had a squiffy eye, David Bowie did rather well for himself.

Bowie blazed a trail as the godfather of glam rock, invented innovative electronic music in Berlin, embarked on forays into funk and 'plastic soul' and, underpinned by an unwavering intellectual inquisitiveness and an extraordinary ability to reinvent himself, continued to break musical and artistic ground right up until his death in 2016.

He was also quite good in *Labyrinth* and *SpongeBob SquarePants*.

One of Bowie's favourite drinks was, unsurprisingly, a bit weird. Schelvispekel, meaning 'haddock brine' is a spicy Dutch liqueur drunk by fishermen. It's produced using brandy and spices but, thankfully, no haddock.

But that still sounds disgusting so, instead, we're bringing it back to Bowie's birthplace with a lovely lager from the Brixton Brewery – whose original home was just a ten-minute walk from Bowie's.

Named after the iconic Brixton street (where huge crowds gathered to party and pay homage on the day of Bowie's death), Coldharbour Lager is a Bohemian-style lager for the ultimate bohemian.

Unfiltered, unpasteurised, unfined (vegan-friendly) as well as beautifully balanced, it's made with two German hops (a nod to Bowie's Berlin days).

9 JANUARY

STEVE JOBS UNVEILS THE FIRST APPLE IPHONE (2007)

GOLDEN DELICIOUS –

On this day back in 2007, the geek inherited the earth when Steve Jobs, chief nerd at Apple, unveiled the iPhone.

Hailed by many as the most significant cultural invention of the 21st century, yet bemoaned by others as a plague on pub conversations, the iPhone has transformed human behaviour in a way that few could have expected at the time.

Still, put your smartphone away and enjoy this delicious drink from Normandy in honour of Steve Jobs: the ultimate 'Norman'.

~

40ml (1¼ fl oz) Calvados Selection | 10ml (½ fl oz) H by Hine cognac
30ml (1 fl oz) aged white port | 10ml (½ fl oz) Poire Williams liqueur
3 dashes of peach bitters | Ice | Apple slice, to garnish

- Pour the ingredients into a mixing glass with ice and stir until chilled. Strain into a chilled coupe and garnish with an apple slice.

10 JANUARY

INDIAN TEA FIRST SOLD IN BRITAIN (1839)

EARL GREY MARTINI –

In Britain, there are very few problems that can't be solved by 'putting the kettle on'. For nearly 400 years, tea has been the cornerstone of British culture and, since usurping ale and gin centuries ago, it remains the nation's favourite drink.

Initially, the tea drunk in Britain came from China. It was a Scottish explorer called Robert Bruce (no, not that one) who first tapped up the tea growers in India's Upper Assam region. After he died in 1824, his younger brother Charles took over and cultivated a nursery of Assamese plants, sending samples to London in 1837.

These Indian tea chests were first auctioned in 1839, and everyone involved celebrated with a lovely cuppa.

OK, it's not an Indian tea, but this posh 'splosh'-infused Martini is marvellous.

~

50ml (1½ fl oz) gin | 35ml (1 fl oz) cold strong Earl Grey tea
20ml (¾ fl oz) lemon juice | 15ml (½ fl oz) sugar syrup
½ free-range egg white | Ice | Twist of lemon peel, to garnish

- Shake ingredients (apart from the garnish) with ice in a cocktail shaker. Double-strain into a Martini glass. Add a lemon twist.

---

11 JANUARY

LAUNCH OF MORSE CODE (1838)

'THREE DOTS AND A DASH' COCKTAIL –

No one uses Morse code anymore. It's been usurped by other communication devices, including the telephone, the internet, carrier pigeons and shouting.

The brainchild of famous portrait painter and brainy boffin Samuel Morse, he came up with the idea after being told of his wife's illness via a messenger on horseback. Sadly, it had taken so long for the letter to reach him that by the time he arrived home, she had not only died, she'd also been buried. Horses, eh?

The code was first publicly demonstrated in front of hundreds of people at Speedwell Ironworks in Morristown, New Jersey. The message they sent read 'Railroad cars just arrived. 345 passengers.'

Everyone clapped.

Created by Don the Beachcomber, the godfather of tiki (see 22 February), the 'Three Dots and a Dash' cocktail is named after the Morse code for 'V' – as in victory!

~

45ml (1½ fl oz) rhum agricole | 15ml (½ fl oz) golden rum
7.5ml (¼ fl oz) dry Curaçao | 7.5ml (¼ fl oz) falernum liqueur
15ml (½ fl oz) lime juice | 15ml (½ fl oz) orange juice
15ml (½ fl oz) honey syrup (2 parts honey, 1 part water)
2 dashes of Angostura bitters | Crushed ice
Pineapple leaves, pineapple spear, an orchid
and three brandied cherries, to garnish

- Briefly shake all the ingredients (except garnishes) in a cocktail shaker with crushed ice. Strain into a tiki mug filled with crushed ice.
- Garnish with the pineapple leaves, pineapple spear, an orchid and three brandied cherries, then serve.

---

12 JANUARY

Kiss a Ginger Day

KING'S GINGER MULLED CIDER –

King Edward VII was a bit of a player who ruffled the feathers of the Royal Household. He spent his time playing golf, shooting animals, inking up his body with tattoos, bedding mistresses, smoking cigars and living the good life in the latest fashions – particularly tweed.

He also liked to muck about in his open-top Daimler. This stressed out his physicians, so they had a liqueur created to keep him warm.

~

50ml (1½ fl oz) The King's Ginger
150ml (5 fl oz) Aspall Draught Cyder | 150ml (5 fl oz) apple juice

- Pour all the ingredients into a saucepan over a low heat and gently warm through. Serve in a handled hot toddy glass.

13 JANUARY

Birth of Marco Pantani (1970)

THE BICICLETTA -

Just five foot seven and weighing 56kg (123lbs), Italian Marco Pantani was one of road cycling's most charismatic, captivating climbers. Yet, much like the mountain stages he adored, his life was full of dramatic ups and downs, and was as triumphant as it was utterly tragic: in 2004, he died from a suspected cocaine overdose after being banned from the sport he loved.

With the ideal featherweight physique for riding bicycles up steep hills at extraordinarily high speeds, his romantic do-or-die riding style – idealistic, unrealistic and favouring intuition over tactics – made him a cult hero.

His swashbuckling style, bandana and earring earned him the nickname El Pirata ('The Pirate'), and he boasted a rich hinterland of extracurricular activities, including painting, writing poetry and serenading women in bars with his guitar.

A statue of Pantani sits at the peak of Plan di Montecampione in Lombardy, northern Italy, where he famously crushed his rivals on the 1541-metre (5055-feet) climb during the 1998 Giro d'Italia. He crossed the finish line with his eyes closed and arms raised to the sky, on his way to winning the whole race.

This part of Italy is also where the Bicicletta cocktail originates. It dates back to the 1930s and is named after the mode of transportation on which its drinkers wobble their way home after overdoing it a bit.

~

Ice | 50ml (1½ fl oz) Campari | 75ml (2½ fl oz) white wine
Soda water | Lemon slice, to garnish

- Build the ingredients in a wine glass over ice and serve garnished with half a lemon slice.

14 JANUARY

Birth of Norman Keith Collins, aka Sailor Jerry (1911)

NORMAN COLLINS COCKTAIL -

Norman Keith Collins, aka Sailor Jerry, was an iconic 'ink' pioneer. Having learned to tattoo as a child while train-hopping across America, Collins honed his techniques in the navy when he was stationed in South East Asia, seeking inspiration from Japanese masters known as Horis, the most gifted and innovative tattoo artists of the time.

After leaving the Navy, Collins settled in Honolulu's infamous Hotel Street, where thousands of American servicemen on shore leave would escape the horrors of war with drink and ink, and by 'sinking the pink.'

During the 1940s and 50s, he deftly dovetailed American and Asian influences to create bold, masculine artwork that had humour and swagger. On the door of his shop he hung a sign saying: 'If you don't think you're man enough to wear a tattoo, don't get one. But don't try to make excuses for yourself by knocking the fellow who does!' It was signed: 'Thank you . . . Sailor Jerry.'

~

50ml (1½ fl oz) Sailor Jerry Spiced Rum | 25ml (¾ fl oz) lemon cordial
25ml (¾ fl oz) lemon juice | Ice | Soda water, to top up
Lemon slice skewered on a cocktail stick
and a maraschino cherry, to garnish

- Shake the rum, cordial and juice in a cocktail shaker with ice, then strain into a Collins glass filled with fresh ice. Top up with soda water, and garnish with a lemon wheel and a maraschino cherry.

15 JANUARY

Boston Molasses Flood (1919)

RUM OLD FASHIONED –

In 1919, a giant steel tank holding 10 million litres of molasses exploded in Boston, unleashing an 8-metre (26-foot) high tsunami of treacle that tore through the city at 35 miles per hour.

Twenty-one people came to a very sticky end, 150 were seriously injured, train tracks were torn apart and houses were destroyed.

All very sad, of course, but it was a long time ago. It's time to move on, with a molasses-fuelled twist on the Old Fashioned.

~

5ml (¼ fl oz) sugar syrup | 2 dashes of Angostura bitters
75ml (2½ fl oz) Diplomático Reserva Exclusiva Rum
Orange zest, for squeezing | Ice

- Place the sugar syrup and bitters in a rocks glass, then add one ice cube and stir. Add about 20ml (¾ fl oz) of the rum and another ice cube, and continue stirring. Keep adding ice cubes and rum, a little at a time, stirring all the while, until all the rum has been added. Stir once more, then squeeze the oil from an orange zest twist over the drink. Drop in the zest, stir and serve.

16 JANUARY

Ivan the Terrible Crowned Tsar of Russia (1547)

VODKA –

Ivan the Terrible really was terrible.

The 16th-century Czar may have expanded Russia's empire and united a divided kingdom, but he wasn't a very nice chap at all.

Ivan unleashed trained bears on monks, chucked puppies from the tops of buildings, and anyone questioning his tyrannical rule would receive a visit from the *oprichniki*, gangs of horsemen dressed in black with an unhealthy penchant for roasting victims alive.

Racked with vodka-fuelled paranoia, Ivan thought his own son was plotting to murder him and take his throne. So, in a fit of rage, he killed him. Turns out, the boy wasn't doing anything of the sort. Ivan should have simply sent him to the 'naughty step'. Or maybe given him a smacked botty.

Unlike the despot after which it is named, 'Ivan The Terrible' vodka is actually quite nice. It's an Osobaya (meaning special) vodka, made in Moscow (just like Ivan), and it goes down easier than a puppy thrown from the top of a building.

---

17 JANUARY

PROHIBITION INTRODUCED (1920)

SOUTHSIDE RICKEY –

Just over a hundred years ago, America introduced Prohibition – a 13-year period of ill-enforced abstention. It was a flipping disaster.

Loads of people died, there was an organised crime epidemic, the government lost a total of $11 billion in alcohol tax revenue – in addition to the $300 million Prohibition cost to enforce – and American society fell apart.

The most famous Prohibition cocktail is the Southside Rickey, an illicit gin drink said to have been enjoyed by Chicago mobsters including Al Capone, whose rough-tasting gin needed disguising with sweetness. (Don't tell him that, though, he'll smash your head in with a baseball bat.)

~

60ml (2 fl oz) gin | 30ml (1 fl oz) lime juice
15ml (½ fl oz) sugar syrup | 6 mint leaves | Ice

- Shake in a cocktail shaker with ice. Strain into an ice-filled tumbler.

18 JANUARY

George Clooney Appointed United Nations Messenger of Peace (2008)

CASAMIGOS ANEJO TEQUILA –

As well we know, being one of the most handsome men in the world is a full-time job. There's the grooming, the moisturising, the daily dips to improve one's core and the regular application of talcum powder around one's dangly bits.

Yet somehow, legendary 'dish' George Clooney still managed to find time for Oscar-winning acting, directing, producing, advertising average coffee *and* performing vital human rights work as a Messenger of Peace for the United Nations.

George ended his role with the UN in 2014, two days after getting engaged to Amal Alamuddin, an internationally acclaimed barrister (he's not the only one who can make coffee) and Human Rights campaigner who, let's face it, is much better qualified to wear the peace-keeping pants in the relationship.

While George remains fully committed to his humanitarian causes, he also released Casamigos, a range of premium tequilas, in 2013. Enjoy the aged, amber-coloured 'anejo' on the rocks.

19 JANUARY

Launch of Tour de France Announced (1903)

COFFEE AND CHAMPAGNE –

The Tour de France was the beautiful brainchild of Henri Desgrange, owner of *L'Auto*, a sports newspaper better known today as *L'Équipe*.

Desperate to steal a march on its sporting rival *Le Velo*, Desgrange announced a six-day race across France consisting of ridiculously long, 17-hour stages.

Riders used alcohol to fuel their pedalling, seeking sustenance from local bars along the route. While his rivals rode on a diet of strong red wine, the first winner, Maurice 'the Chimney Sweep' Garin took a different approach, once claiming that a combination of coffee and champagne gave him the legs for victory.

We suggest you drink them separately, one chasing the other.

JANUARY 20

BARACK OBAMA INAUGURATION ADDRESS (2009)

SOFIE, GOOSE ISLAND –

More than 1 million expectant Americans gathered on Washington's Capitol Hill to hear the inauguration address of the nation's first African-American president.

Echoing the transcendental principles laid down by Abraham Lincoln, evoking the spirit of JFK and delivering his words with the laconic lilt of Martin Luther King, Barack Obama gave a passionate, poignant yet pragmatic sermon.

Obama spoke eloquently of sacrifice and suffering and, crucially, he spoke to Americans like adults: 'We remain a young nation, but in the words of scripture, the time has come to set aside childish things.'

Nine years later, Donald Trump, an actual child in a man's body, was elected President.

If that leaves a rather unpleasantly sour taste, we suggest you replace it with a much more pleasant one. Chicago brewery Goose Island's Sofie is a superb *saison*-style beer and an ideal stepping stone into the world of sours.

21 JANUARY

THE MYSTERIOUS AFFAIR AT STYLES (THE FIRST HERCULE POIROT NOVEL) RELEASED (1920)

HERCULE STOUT, ELLEZELLOISE BREWERY –

With 2 billion sales under her belt, only God (who famously wrote the Bible) and William Shakespeare have shifted more books than Agatha Christie. Hercule Poirot, her brilliant, whisker-twiddling Belgian sleuth is loved by readers, yet famously loathed by Christie, who described him as a 'detestable, bombastic, tiresome, egocentric little creep'.

Played by an esteemed list of actors over the years, including John Malkovich and Orson Welles, the brainy little Belgian is synonymous with David Suchet, who perfected the detective's distinctive short-strided walk by placing a coin in the crack between his butt-cheeks – a 'trick' apparently learned from Laurence Olivier.

Christie was deliberately vague about Poirot's origins, but his accepted birthplace is the town of Ellezelles, which is also home to the excellent Ellezelloise Brewery and their classic Hercule Stout: a smooth, strapping, velvet-jacketed vortex of caramel, cocoa and espresso. At 9 per cent ABV, a couple of these will, like Hercule himself, solve everything.

22 JANUARY

DEATH OF QUEEN VICTORIA (1901)

ROYAL SOUR –

The reign of Queen Victoria coincided with arguably the most magnificent period of British history.

With Victoria 'in charge', Britain pretty much ran the global show. Its economy and industry were the envy of the world, and the Brits even beat Australia at cricket on a relatively regular basis.

Despite her diminutive four-foot-eleven stature, Queen Victoria was an eminent imbiber of alcohol, with her preferred poison being an unusual mix of whisky and red wine. Together. In the same glass.

She was particularly partial to Claret and Vin Mariani, a drink made by Angelo Mariani by steeping cocoa leaves in French red wine for six months. Alleged to be the original recipe for Coca-Cola, each fluid ounce contained 7.2 milligrams of cocaine.

Today's drink is cocaine-free. We've replaced the bourbon in the Royal Sour, an enduring Claret-clad classic cocktail, with Royal Lochnagar 12-Year-Old, a light, fresh highland Scotch famed for its royal warrant (issued by Queen Victoria – see 10 February).

~

50ml (1½ fl oz) Royal Lochnagar 12-Year-Old
15ml (½ fl oz) lemon juice | 15ml (½ fl oz) sugar syrup
Dash of Angostura bitters | Ice | 15ml (½ fl oz) Claret

- Shake all the ingredients (apart from the Claret) in a cocktail shaker with ice. Strain into a large tumbler brimming with ice. Trickle the red wine around the surface of the drink. For authenticity, drop a bit of Coke on the top, too.

---

23 JANUARY

BLUEST DAY OF THE YEAR

BLUE HAWAIIAN –

Today is statistically the day when people in the Western world feel the most blue.

But instead of buying into this collective 'boohoo', wipe away the snot bubbles from your nostrils, slip into some *Magnum P.I.*-style budgie-smugglers or a bikini, put on some Mungo Jerry and channel the tiki bars of Hawaii with a kitsch old-school classic blue drink.

Best enjoyed while basking in the glare of a SAD lamp.

~

30ml (1 fl oz) Bacardi Carta Blanca light rum
30ml (1 fl oz) Bols Blue Curaçao | 90ml (3 fl oz) pineapple juice
2 tablespoons cream of coconut | 5ml (¼ fl oz) lime juice | Ice

- Shake all the ingredients in a cocktail shaker with ice, then strain into a hurricane glass filled with ice.

---

24 JANUARY

National Beer Can Day (USA)

FELINFOEL IPA –

The beer can was once viewed as the village idiot of vessels, its head pompously patted by haughty craft-beer connoisseurs like a grinning smudge-faced simpleton who points at planes.

But now, cans are the regarded by these same connoisseurs as ideal containers in which to house their hooch. There are several reasons for this: cans don't break, they're lighter than bottles, they're more environmentally friendly and they're easier to transport. But, more crucially, the can keeps beer consistently fresher and more flavoursome for longer, protecting beer from its two arch enemies: sunlight and oxygen.

Back in 1935, Krueger's of New Jersey officially produced the first canned beer, but the Felinfoel Brewery in South Wales – a brewery with family interests in the local tinplate industry – followed suit 11 months later by packaging their brews in tin.

The original cans are now collector's items, and Felinfoel now celebrates more than 85 years of canning with an awesome American-style IPA using British hops and Welsh water.

25 JANUARY

Burn's Night

HARVIESTOUN OLA DUBH –

It's Burns Night.

In any other month of the year, no one would countenance an evening of unintelligible Scottish poetry, blaring bagpipes that sounded like a fire in a pet shop, and a 'supper' of root vegetables and a stomach stuffed with suet, salt and various bits of sheep, all tied up with a dress code strictly stipulating 'no underpants'.

But in January? We're all over it. What time does it start?

Salute the promiscuous Scottish poet with Ola Dubh ('black oil'), a delicious dark beer aged in barrels formerly occupied by Highland Park 12-Year-Old whisky.

26 JANUARY

Rum Rebellion (1808)

HUSK PURE CANE AUSTRALIAN RUM –

When rum was first exported to Australia, it was used as currency to finance buildings and catch criminals – you could even buy a wife with a gallon of the stuff.

The exchange of rum and other spirits fuelled the nation's embryonic economy until the arrival from London, of a new governor: naval officer William Bligh, infamous as the commander of HMS *Bounty* during its mutiny.

When Bligh, a professional pain in the backside, belligerently decided to ban alcohol as payments for trades, it seriously pissed off some prominent locals, none more so than Major George Johnson and John Macarthur, who were both making bundles of cash from booze-driven business.

In 1807, with tensions mounting between the two, Bligh ordered Macarthur's arrest over a shipping misdemeanour. But when he was released on bail, Macarthur and Johnson led a regiment of the New South Wales Corps to Bligh's house, had him arrested and ruled the colony for the next two years.

It remains the only military coup in Australia's history.

27 JANUARY

BIRTH OF MOZART (1756)

MOZART MARTINI –

Johannes Chrysostomus Wolfgangus Theophilus Mozart was, by all accounts, an annoyingly talented child.

By the age of ten, while his friends were drawing stick figures, Mozart had written 16 sonatas. And they weren't crap ones either. By the age of 15, he had 20 symphonies to his name. During his short life, he wrote *Don Giovanni*, *The Marriage of Figaro*, *The Magic Flute*, *Così Fan Tutte*, numerous late piano and woodwind concertos and, most famously, his Requiem, written when he knew he was dying.

He didn't, however, write 'Twinkle, Twinkle, Little Star' – a myth propagated by fools.

Sourced from his native city of Salzburg, Mozart Dark Chocolate Liqueur is magnificent in an uber-indulgent chocolate Martini. It's definitely not for kids – not even precocious piano-playing ones.

~

50ml (1½ fl oz) gin
15g (½ oz) dark chocolate (70 per cent cocoa solids), melted
25ml (¾ fl oz) Mozart Dark Chocolate liqueur | Crushed ice
Raspberry, to garnish

- In a jug, whisk together the gin and melted chocolate until smooth. Add the Mozart chocolate liqueur. Pour this mixture into a cocktail shaker with crushed ice. Shake, then strain into a Martini glass. Garnish with a raspberry.

28 JANUARY

DEATH OF HENRY VIII (1547)

TYNT MEADOW TRAPPIST ALE –

Before Henry VIII's reign, monks pretty much brewed all the beer in Britain, but the misogynistic monarch famously put an end to that during the Reformation.

Unlike Jay Z, Henry VIII had lots of problems directly, or indirectly, related to the women in his life, and these issues forced him to dissolve all the monasteries in the 16th century.

With no monks around to work magic with their mash-forks, British brewing passed into the secular hands of landowners and farmers – and there it stayed until 2017, when the Mount St Bernard Abbey in Leicestershire started brewing England's first Trappist beer (a beer brewed by monks in a monastery for good causes).

Established in 1835, Mount St Bernard Abbey is England's only Cistercian abbey. It once operated as a brewery, and more recently, the monks had made money from their dairy farm – but, you know, milk simply isn't the cash cow it once was. So, in order to continue their charitable 'monk-y' business, they began brewing Tynt Meadow, a delicious, full-bodied, bottle-conditioned Dubbel. Mahogany in hue and beige of head, it's full of dark fruit, chocolate, liquorice and winter spice flavours, with a lovely nutty finish.

29 JANUARY

JAMES JAMERSON BORN (1936)

METAXA 12 STARS BRANDY –

James Jamerson was a phenomenally gifted bass player who is widely considered as a founding father of funk.

His impressive bass lines propelled the renowned sound of the legendary Motown record label and its illustrious Funk Brothers rhythm section, the backing band of Motown stars including Stevie Wonder, Diana Ross, Smokey Robinson and Bill Withers.

When Marvin Gaye was recording his iconic album, *What's Going On*, he insisted on having Jamerson play bass. When no one could get hold of him, Gaye trawled the bars of Detroit until he discovered him, slightly worse for wear, in a local club.

Gaye brought Jamerson back to the studio, but he was so inebriated that he was unable to sit upright in the chair. So Jamerson got down on the floor and, drunk out of his mind and lying flat on his back with his eyes closed, he unleashed one of the most beautifully laid-back bass lines ever recorded.

Rather than dilute his dexterity, Jamerson claimed drink loosened up his playing, and his tolerance for alcohol was legendary. When asked why he drank so much, he answered 'Because I like the taste of it.'

Jamerson always kept a bottle of Metaxa 12 Stars Greek brandy in his bass case: a blend of double pot-stilled brandy matured for a minimum of 12 years in Limousin oak and then blended with Muscat wines and an infusion of rose petals and Mediterranean herbs.

---

30 JANUARY

UK LEAVES THE EU (2020)

CANTILLON LAMBIC –

What better way to commemorate the UK's relations with the European Union than with a slightly sour beer style, brewed in an unremarkable corner of Brussels for centuries?

More a museum than a microbrewery, Cantillon is a tumbledown time warp. Big frothing oak barrels line its crumbling walls and soar high into the musty, dusty rafters; fusty fumes fill the nostrils; pipes drip, hoses trip and decrepit 19th-century copper vessels and wooden tuns clunk, churn and creak.

Its beer, fermented using wild yeast in the air, is funky and farmyard-y and adored by beer aficionados, but, much like the EU, it can really polarise opinion. Some regard it as the most quixotic and cultured expression of the brewing art, while others simply can't see the attraction of a sour, tart beer that tastes like a goat smells. We like it though.

---

31 JANUARY

Death of A. A. Milne (1956)

HIVER BEER –

In honour of the creator of Winnie-the-Pooh, whose love of honey is renowned, we're having Hiver Blonde, a glorious golden ale made using three different honeys from British beekeepers.

Bees, it turns out, are fascinating little fellows. They won't sting you unless they feel under threat, and they'll give you a warning first by raising one of their legs. When they do sting you, their arse falls off. They have five eyes (which see 300 frames a second), and when flowers 'hear' a bee's wings (which move at a rate of 230 times a second) they make their nectar sweeter. A third of all food we eat is pollinated by bees. And, best of all, bumblebees' penises explode when they ejaculate.

Honey, meanwhile, is better for treating coughs and colds than antibiotics and many over-the-counter medicines. However, it won't protect you from the most dangerous bee in the world: Hepatitis B.

FEBRUARY

1 FEBRUARY

Publication of Lord Byron's *The Corsair* (1814)

BORDEAUX –

A warm welcome back to any January abstainers. Remember, if you drink less but better for the rest of the year, you won't *need* to give it up for an entire month next January. So, let's all flip abstinence the bird by turning to a pioneer of overindulgence, Lord George Gordon Byron.

While not crafting celebrated poetry, Lord Byron dabbled in incest, drank out of the skulls of (dead) monks, kept a pet bear at university and had a sexual appetite that landed him all manner of loin lurgies by the time he was 21. With antics to make today's cautious 'Insta-influencers' look like total lightweights, he amassed an army of followers. When *The Corsair* was published in 1814, it sold 10,000 copies in the first 24 hours.

Despite his debauched day-to-day, though, Lord Byron was discerning in his drinking choices and opted for 'fine Claret' or light Bordeaux wines. Legend has it that in 1821, he visited the Bordeaux estate Château Chasse-Spleen, where he remarked of the wine: '*Quel remède pour chasser le spleen*,' giving the château its name.

Today's Chasse-Spleens would no doubt please the poet: they are as deep, warm, spicy and full-bodied as his explorations into his fellow humans.

2 FEBRUARY

Publication of *Ulysses* by James Joyce (1922)

JAMESON WHISKEY –

James Joyce was one of the greatest novelists of the 20th century and the godfather of modernist literature, but it was his affection for whiskey and the pub that earns him our respect. And while scholars of *Ulysses* focus on its echoes of Homer's *Odyssey*, we prefer to consider how, rather than fighting angry gods and monsters returning from the Trojan war, Joyce's protagonist, Leopold Bloom, spent his days making breakfast, feeding the cat, having a wank and, crucially, going to the pub.

Moreover, Joyce's novel is a passionate defence of buying rounds in pubs. While Joyce was writing, Ireland was coming under increasing pressure from its British oppressors and uptight teetotallers to ban the tradition of 'treating' – getting a round in. In his book, Joyce hails 'round-buying' as a symbol of Irish autonomy and resistance to British rule. Essentially, he used *Ulysses* to thumb his nose at the temperance movement.

Joyce was a committed barfly, claiming he was at his most creative when in his cups – and it was Jameson Irish Whiskey that he poured into them. Jameson inflamed his imagination and blew away the cobwebs of literary convention. While it also made some of his writing incomprehensible to us (*Finnegan's Wake* is a real ball-ache to get through), his novels were soaked in references to his favourite tipple: John Jameson even gets a name check in *Ulysses*.

A triple-distilled blended whiskey, Jameson is produced at the Midleton Distillery in Cork, home to a host of other exceptional blends and single malts. Get a round of them in for Joyce today.

3 FEBRUARY

FIRST WINTER OLYMPICS IN CHAMONIX (1924)

SNOWBALL FIGHT COCKTAIL –

The Alps is rich in discerning drinking options, from botanical spirits like chartreuse or Dolin Génépy, to wines and craft beers. Even so, we mark this moment with a cocktail using Kahlúa.

~

Ice | 15ml (½ fl oz) Absolut Vanilla
15ml (½ fl oz) Kahlúa mint mocha liqueur | 1 tablespoon single cream
Dash of Pernod absinthe (optional) | Mint sprig, to garnish

- Place the ice in a rocks glass, then pour over the vanilla vodka, Kahlúa, cream and absinthe (if using). Stir, then serve, garnished with a sprig of mint.

4 FEBRUARY

SS POLITICIAN SETS SAIL (1941)

TALISKER WHISKY –

When the cargo ship SS *Politician* left Liverpool for America in 1941, it transported, among other less important bits and bobs, 260,00 bottles of whisky. Two days into the voyage, though, a storm forced it aground off the Hebridean Isle of Eriskay. Learning of the wreck, the crafty locals commandeered tens of thousands of bottles, claiming what they believed to be a legitimate bounty haul under marine salvage law. Sadly, spoilsport customs officer Charles McColl didn't agree, and prosecuted a number of these heroes for thieving. What a bell-end.

The wreck remained, though, and in 1987 Donald MacPhee, a

local South Uist man, recovered eight more bottles, selling them at auction for £4,000 ($5,500). A year later, a pub opened, named 'Am Politician' (The Politician) in honour of the story. The entire carry-on inspired Compton Mackenzie's book *Whisky Galore*, and subsequently an Ealing Studio comedy.

Excellent work, everyone – except McColl, who, if you remember, is a bell-end.

Say, 'In your face, tax man,' as you drink Talisker, a (legal) Inner Hebridean whisky from Skye.

5 FEBRUARY

AENEAS COFFEY RECEIVES PATENT ON HIS STILL (1831)

IRISH WHISKEY OR SCOTTISH BLENDED WHISKY –

It's hard to believe Irish whiskey was once one of the bestselling spirits in the world, but it was. So, er, believe it. History records the Irish as the original whiskey distillers, and by the early 19th century, the popularity of their spirit far outstripped that of Scotch. Were it not for a series of bad choices and a dose of bad luck, it might remain top dog today.

English oppression and US prohibition were among the challenges faced by the Irish, but it was their decision to stick with traditional pot-distilling techniques that truly did for them. By the early 19th century, column-distillation technology was modernising the spirits industry, with new, shiny taller stills enabling a continuous process that produced highly rectified, 'clean' spirits. Among the pioneers was an Irish excise man turned engineer named Aeneas Coffey, who received a patent on his own flashy column still concept in 1831.

But when presenting the Coffey Still to his compatriots, Aeneas found them reluctant. Despite the fact Coffey's tech could distil a crisper spirit at a faster rate and for less money, the traditionalists in Ireland preferred to stick with a more robust and flavoursome pot-still distillate. Not to be deterred, Coffey took his idea to the Scots, who, understanding they could use it to make their spirit cheaper, obviously embraced it.

The Coffey still eventually became a core tool in the production of grain whisky in Scotland, and, when mixed with single malt, a new, easy-drinking blended Scotch became the most popular whisky in the world. Thus, the Scots used the Irishman's Coffey still to help make whisky-without-the-'e' a global success, decimating the Irish whiskey industry in the process. And if you've ever seen a performance of *Riverdance*, you'll know the Irish have been kicking themselves ever since (see 9 February).

Feel free to celebrate Aeneas today with his native Irish whiskey (or, indeed, a Scottish blend).

---

6 FEBRUARY

BIRTH OF BOB MARLEY (1945)

JAM JAR HAND SHANDY –

Enjoy a cocktail with jam in. Because, you know, we hope you like jammin' too? As in the song, right? *We're jammin'*?

No?

Oh.

Anyway, Tim Fitz-Gibbon created this drink at Raoul's Bar in Oxford, England.

~

Crushed ice | 50ml (1½ fl oz) Woodford Reserve bourbon
25ml (¾ fl oz) lemon juice | 2 teaspoons strawberry jam
1 teaspoon vanilla-infused sugar | Soda water, to top up
Strawberry, to garnish

- Fill an empty jam jar with crushed ice. Shake the bourbon, lemon juice, jam and sugar in a cocktail shaker to break up the jam and combine. Strain over the ice, top up with soda water, then stir and top with more crushed ice. Garnish with a strawberry and serve.

7 FEBRUARY

A PRAT DESTROYS THE PORTLAND VASE (1845)

DEAD GUY ALE, ROGUE –

Never has our 'drink less, drink better' mantra been more applicable than on this day in 1845, when a drunk visited the British Museum, chucked a precious artefact into a case full of more precious artefacts, and subsequently destroyed the priceless Portland Vase.

The vase was famously adorned with images of naked people, and was considered a masterpiece of Roman glass, dating back 2,000 years. It was owned by William Henry Cavendish Cavendish-Bentinck, 3rd Duke of Portland, who miraculously did not press charges.

The drunken rogue, who had apparently been on the sauce for days, was named as William Lloyd – although it later transpired his real name was William Mulcahy. He had changed his name while in hiding from his family in Dublin. In our experience, there are far better ways to keep a low profile than bunging artefacts about in central London museums.

Incredibly, the vase has been restored and is still on display, so celebrate this happy ending with a craft beer from Portland, USA, a city that boasts many brilliant craft-brewing legacies. Some of the best exponents happen to be served in Mary's Club, a strip club in the city.

'Why bring up a strip club?' we hear you ask. Well, 'bare' with us, because it is our understanding, from people who have been to Mary's (which are not us), that this historic Portland site prides itself on an exceptional craft beer menu. So: Mary's is in Portland (Portland Vase); it serves the Oregon craft beer Rogue (William);

and you will see naked people here (Portland Vase again).

Rogue started life as initially a brewpub in Ashland, just south of Portland, and the brewery team has pioneered adventurous styles of beer. We recommend their Dead Guy Ale, because there are loads of stories about dead guys in the British Museum, and, when he smashed the Portland Vase, we imagine William Mulcahy's mates said something along the lines of: 'You're well dead, guy.'

The beer is inspired by German maibocks, so it's a bit nutty and slightly sweet, with some lovely dried and red fruit notes, before leaving a bitter finish – similar to the taste left in the mouth of the museum curator.

---

8 FEBRUARY

JACK NICHOLSON ATTACKS A CAR WITH A GOLF CLUB (1994)

JOHN DALY COCKTAIL –

When the great Jack Nicholson attacked a car with a golf club during an episode of road rage in 1994 – an event he apologised profusely for – he claimed he selected the 2-iron because it was a club he didn't use on the course. For anyone who shares his passion for golf clubs, why not enjoy a John Daly, named after the 1991 PGA champion?

~

50ml (1½ fl oz) vodka | 15ml (½ fl oz) Grand Marnier
45ml (1½ fl oz) sweetened cold English breakfast tea | Ice
Lemonade, to top up | Lemon slice, to garnish

- Shake the vodka, Grand Marnier and tea in a cocktail shaker with ice, then strain into a highball glass full of ice. Top up with lemonade, then serve, garnished with lemon slice.

9 FEBRUARY

RIVERDANCE PREMIERE (1995)

SAUTERNES –

Step up (and down and up again) Michael Flatley, who led the premiere of *Riverdance* at the Point Theatre in Dublin, Ireland, in 1995. In an interview with the *Irish Times*, Michael suggested he was once partial to a glass of Sauternes, and this sweet Bordeaux white wine adds a lovely little kick to a dessert dish.

10 FEBRUARY

WEDDING OF QUEEN VICTORIA (1840)

ROYAL LOCHNAGAR WHISKY –

Who doesn't love a Royal wedding? Really brings a nation together. Right, guys?

Our favourite of all time was probably the nuptials of Queen Victoria and Albert of Saxe-Coburg and Gotha in 1840 – remember that one? Belter.

Queen Victoria loved food and drink, and once enjoyed a dram at the diminutive New Lochnagar whisky distillery, which sits on the edge of the Balmoral Estate. Such was the impact of her visit, the distillery was renamed Royal Lochnagar and still makes exceptional single malt whisky, characterised by its rich, malty, spicy profile, with notes of winter fruit. (For more on Queen Victoria and Lochnagar, see 22 January).

11 FEBRUARY

NATIONAL FOUNDATION DAY (JAPAN)

HIBIKI SUNTORY WHISKY –

The Japanese sip
On this proud Foundation Day
Whisky of Japan

(For any thickos reading, this is a haiku. And, I think we can all agree, it's a really good one.)

12 FEBRUARY

BIRTH OF CHARLES DARWIN (1809)

FLORIS MANGO, HUYGHE BREWERY –

Darwin's theory of evolution, as explored in his *On the Origin of Species*, might be advanced a little with a footnote on beer.

Scientists have asserted that fruit was naturally fermenting long before humans evolved, so it's now accepted (by us) that boozy banana aromas rising into the treetops would have inspired primates to leap from the leaves to the forest floor in search of these sweet-smelling, intoxicating treats.

This fermenting fruit would've given the primates a buzz, maybe made them a little amorous, but was also easier to digest so essential calories became more available – plus it had the benefit of alcoholic antibodies to fight off infection. Thus, it was the swiftest and strongest swingers who made it to nature's open bar first, and they who benefited the most. Ergo, booze was the instigator for survival of the fittest.

It's not just us saying this: Nathaniel Dominy, a biological anthropologist at Dartmouth College, was quoted saying much the same thing in a feature in the *National Geographic*. So, while we're paraphrasing a bit, humans are genetically predisposed to drink, thanks to our ancestors' love of a boozy fruit. Actual science.

We recommend Floris Mango, a wheat beer brewed with the addition of mango juice made by the Belgian Huyghe Brewery. We could've suggested Huyghe's banana beer here, but it felt a little on the nose. Besides, we've used that for a story about the reintroduction of bananas on 25 February.

13 FEBRUARY

DEATH OF DR ADAMS (BUT NOT FROM A HANGOVER) (2019)

WARNER'S 0% BOTANIC GARDEN SPIRIT

The only guaranteed way to avoid a hangover is to drink less, but drink better, as per our mantra, but should you happen to awake with rhythmic thumps behind your eyes and the taste of a sewage worker's shoelace on your tongue, you'll undoubtedly pop a few painkillers to try to compensate.

So, let us raise a glass of non-alcoholic spirit today to Dr Stewart Adams, OBE, the pharmacist whose commitment to banishing a 'big head' was so consuming that he took an idea from his own throbbing bonce and invented ibuprofen.

Intent on perfecting his remedy, Dr Adams tested it on himself after a big night out. Forced to deliver an important speech while perhaps fighting back Jägerbomb bile, he dosed up on his new painkiller pills and delivered his address emphatically.

Dr Adams reached the rare old age of 95, and while he sadly died on this day in 2019, it was not from a hangover.

14 FEBRUARY

Valentine's Day

LONDON PORTER, MEANTIME –

St Valentine is rearing his chubby, cherubic head again and, regardless of your relationship status, you'll no doubt wish he wasn't. A stilted, sterile meal awaits those couples who've been together a while, bereft of spontaneity and with nothing but the awkward clinking of knives and forks to fill the void of natural conversation.

But if you are determined to try to bring romance to the table today, we suggest an Argentinian red.

As Feargal Sharkey will tell you, a good heart is hard to find these days so when you do discover one, look after it. Argentinian wine, made with grapes grown in hot yet semi-moist climates, is rich in flavonoids which can help reduce the levels of low-density cholesterols in the body. It makes gummy platelets in the arteries less sticky so that they don't get trapped, form blood clots and result in an *actual* broken heart.

15 FEBRUARY

Birth of Galileo (1564)

ITALIAN WINE –

Galileo Galilei reportedly once said: 'Wine is sunlight, held together by water.' Which is nice, but while he wasn't busy providing hackneyed quotes for tea towels, the astronomer's other achievements include the invention of a new telescope and the discovery of Jupiter's four moons.

Discovering one moon is one thing (it's *one* moon, after all), but discovering four? Well, he must've been over the moons. As a side note, we'd appreciate some credit for assisting in the ongoing search for new moons during our adolescent years, mostly from the back window of our school bus.

---

16 FEBRUARY

National Almond Day (USA)

AMARETTO SOUR –

An entire day honouring the almond: it's nuts, right? But rather than nuts, the almond tree actually bears a drupe in the form of big, juicy plums, inside which are the seeds that we erroneously refer to as nuts.

When it comes to beer, most talented brewers rarely get drupe confused with their nuts, and have used almonds in styles such as stouts. But more familiar to almond fans will be the spirit amaretto. In further updates for pedants, the brand most associated with amaretto is Disaronno, which actually acquires its own nutty flavour from apricot husks rather than the almond nuts (seeds). Either way, it still works in an Amaretto Sour.

~

50ml (1½ fl oz) Disaronno
25ml (¾ fl oz) lemon juice (freshly squeezed)
1 dash Angostura bitters | 15ml (½ fl oz) egg white
Ice | Cherry, to garnish

- Shake the Disaronno, lemon juice and egg white in a cocktail shaker. Add some ice and shake again, then strain into a rocks glass over some more ice. Garnish with a cherry and serve.

17 FEBRUARY

DEATH OF THELONIOUS MONK (1982)

BROTHER THELONIOUS,
NORTH COAST BREWING CO. –

Squibbel-dee-doo-wop, doo-bee-pop: jazz. Thelonious Monk was one of the greatest musicians of the 20th century, and a pioneering pianist in the craft of improvisational jazz. Among the jazz man's fans are North Coast Brewing Co., creators of Brother Thelonious beer, a drinkable, deftly balanced Belgian-style abbey ale. This robust, mahogany-hued homage to the jazz legend hits high notes of chocolate, chipotle and a cinnamon spice. Beer sales also help support the jazz education programmes of the Monterey Jazz Festival. Hmm. Nice.

18 FEBRUARY

HUNTING WITH HOUNDS BANNED (2005)

FOXHOLE GIN –

As the 'hear, hear' barks of MPs carried from Westminster to the Home Counties in 2005, fantastic country foxes screamed the last of their weird, hauntingly high-pitched love-making noises into the night, rolled over and rested a little easier in their dens. They woke to discover hunting would continue without dogs, though, so if you side with this bushy-tailed omnivorous mammal, put pressure on the politicians, for fox sake.

All of which leads us seamlessly to Foxhole Gin, which is blended with a distilled wine, otherwise known as a eau de vie, made using surplus wine grapes from the Bolney Wine Estate in Sussex, and, in keeping with a more humane stance on hunting, no foxes are used in the making of it. The gin builds on the use of this eau de vie with a more traditional botanical bill of juniper, coriander, angelica seed, orris root, liquorice root, bitter orange, fresh lemon and grapefruit zest.

19 FEBRUARY

CLINTON'S BUDGET ENFORCEMENT AND DEFICIT REDUCTION ACT (1993)

THREE-MARTINI LUNCH -

President Bill Clinton left the Oval office with a smoke trail of controversies behind him, but his most damaging political legacy was the menacing manacle he applied to the three-martini lunch. Granted, he injected an exceptional 4 per cent economic growth into the ledgers, so perhaps he knew what he was doing, but let's not blow smoke up his ass. We, as leading global economics experts, argue that had he not signed the Budget Enforcement and Deficit Reduction Act of 1993, which impacted those luncheon martinis, that growth could have been 5 per cent.

The three-martini lunch was originally an imbibing institution that encouraged 1950s businessmen to enjoy three martinis over a working midday meal, the theory being that a little juice could stimulate the warriors of Wall Street and Mad Men of Madison Avenue and inspire the kind of deals our current bunch of business bum clowns can only dream of. Crucially, the martinis were 100 per cent tax deductible.

Jimmy Carter tried to stymie the tax break, claiming the working classes were subsidising the indulgences of executives, but it wasn't until 1986 that the first tax code reduced the break from 100 per cent to 80 per cent, thus dubbing the indulgence a 'two-martini lunch'.

This was bad enough, but then came Clinton. As impenetrably dull as it otherwise seemed, his deficit reduction bill cut the tax break back to whopping 50 per cent on a business-related meal. Did this result in an estimated savings to the government of $15.3 billion over five years? Well, yes, it did . . . But, we cry, at what cost to creative business endeavour?

It has always been our assertion that a martini can unlock the shackles of an imprisoned imagination, particularly when it comes to radical musings on otherwise *well* boring financial matters. Admittedly, drinking less but better will help avoid getting out of your box while you try to navigate your inbox, but with a little relaxation of inhibitions, there are no boundaries on what we can achieve. Perhaps opt for one instead of three, but when you do, we recommend a classic gin, stirred, with a modest dash of vermouth and strained into a chilled cocktail glass.

---

20 FEBRUARY

Vincent Van Gogh moves to Arles (1888)

ABSINTHE WITH A FEW DROPS OF WATER –

Henri de Toulouse-Lautrec walks into a Parisian bar and sees Vincent Van Gogh in the corner having a drink.

'Hello Vincent,' he says. 'Do you want an absinthe?'

'No thanks, Henri,' replies Van Gogh. 'I've got one ear.'

We done a funny. You're welcome.

Van Gogh moved to Arles in 1888 to focus on his art and take a break from the boozy Paris lifestyle, but he quickly discovered absinthe was more available in the south. Arles was renowned for both its abundance of natural sunlight and, rather ominously, an absinthe consumption four times greater than the national average, ensuring the artist's relationship with the green fairy became excessive and infamous.

But while the deterioration of Van Gogh's mental health has been blamed on his dalliances with the absinthe, his increasingly bizarre behaviour – the eating of paint and cutting off his left ear to give to a prostitute – were part of a more serious mental health condition. And while absinthe is forever associated with wild behaviour, the spirit is no worse for you than any other drink – in moderation.

Nonetheless, Van Gogh's affair with absinthe bizarrely continued beyond his death in 1870. He was buried in a cemetery in Auvers, and an ornamental Thuja tree was planted next to the grave. The tree is a rich source of thujone, the hallucinogenic chemical also found in wormwood and used to make absinthe.

When they moved Van Gogh's coffin 15 years later, so he could be re-buried next to his brother Theo, they discovered that the roots of the tree had spookily wrapped themselves tightly round the casket.

---

21 FEBRUARY

KARL MARX AND FRIEDRICH ENGELS PUBLISH *THE COMMUNIST MANIFESTO* (1848)

T.E.A, HOGSBACK BREWERY –

Regardless of your politics, it's important to stress that Marx and Engels conceived their historic theories in pubs. The Red Dragon in Manchester and the Museum Tavern in London are among those laying claim to their patronage, with the lefty liquor lovers seen frequenting both boozers and nursing pints while conferring on capitalist corruption.

So, raise a glass to the pair today, and make it a pint from the Hogsback Brewery in Surrey, England, where they use local ingredients, including hops from their very own hop garden. Among the brews is best bitter T.E.A., which we suspect Marx would prefer to splosh, because, as he once said, proper tea is theft.

22 FEBRUARY

Birth of Ernest Gantt's (1907)

COBRA FANG –

Known to the many as 'Don the Beachcomber', or 'Donn Beach', Gantt is widely recognised as a founding father of tiki drinks. His Polynesian-themed bar in Hollywood became a favourite with the famous faces of the Sunset Strip, who would stay longer than intended when unexpected LA rain showers pelted the bar's windows. These were invariably controlled by Don, who sneakily set up a sprinkler on the roof.

In his honour, try the Cobra Fang, with the added bite of overproof rum.

~

Ice | 15ml (½ fl oz) aged Jamaican rum | 15ml (½ fl oz) overproof rum
15ml (½ fl oz) falernum liqueur | 15ml (½ fl oz) fresh lime juice
15ml (½ fl oz) fresh orange juice | Dash of grenadine
Dash of Angostura bitters

- Fill a blender with ice cubes. Add all ingredients, then blend. Pour into a highball or wine glass and serve.

23 FEBRUARY

Death of Stan Laurel (1965)

ANOTHER FINE MESS PALE ALE,
ULVERSTON BREWING COMPANY –

Stan Laurel finally found a fine mess he couldn't get out of in 1965, dying a few days after a heart attack. At his funeral, Buster Keaton described him as the funniest of all the Hollywood comic greats of his era. And, Buster will be pleased to hear, we agree.

Stan was born in Ulverston, Cumbria, now home to the Ulverston

Brewing Company, which makes a sweet and lightly citrusy pale ale in the comic's honour called, rather conveniently, Another Fine Mess.

Incidentally, we almost got ourselves into another fine mess by duplicating this entry on 7 August to mark the death of Oliver Hardy – in our defence, there are a lot of dates to cover here.

24 FEBRUARY

FIDEL CASTRO STEPS DOWN (2008)

EMINENTE CUBAN RUM –

Fidel Castro managed 49 years as Communist leader of Cuba, making him the longest non-royal head of state in the 20th century – so, well done Fidel. He also had a beard. So highly regarded was his facial hair by his people, that when the CIA were considering ways to dislodge the disruptive leader, they devised a plot to make his beard fall out using a chemical depilatory absorbed via his feet. If you think just bumping him off sounds easier, then note there were more than 600 failed attempts, so perhaps not.

Despite chasing the Bacardi family out of Cuba when he came to power, Fidel did champion local rum, and might've enjoyed Eminente, a light on its toes, fruity but sweet seven-year-old spirit, which comes in a beautiful bottle boasting a glass crocodile-skin texture.

25 FEBRUARY

BANANAS RETURN TO THE UK (1945)

MONGOZO BANANA, HUYGHE BREWERY –

Ooh, have a banana. They are packed with potassium power, while also delivering 15 per cent of your daily vitamin C requirement – plus, they come in a great bendy shape with a comedy slippery skin. No surprise then, that the Brits were devastated when the national supply was cut off during the Second World War. So sorely missed were bananas that Harry Roy enjoyed a wartime hit song titled 'When Can I Have a Banana Again?'. The answer, we now know, turned out to be 25 February 1945.

In honour of this date, why not go bananas over the Mongozo Banana, a banana-flavoured Belgian wheat beer from Huyghe Brewery. Huyghe Brewery based in eastern Flanders, near Ghent, and is celebrated for its Delirium Tremens Belgian Strong Ale. Mongozo Banana is part of their exotic fruit beer range (another of them is featured on 12 February). Far from being a novelty, it is gluten-free and supports Fairtrade and sustainable habitats. Brewed with orange peel and coriander, then blended with banana juice, it is inspired by a traditional recipe from the Masai people of Kenya known as *urwaga*.

---

26 FEBRUARY

BIRTH OF VICTOR HUGO (1802)

BLENDED SCOTCH –

While the French novelist created classics *Les Misérables* and *The Hunchback of Notre-Dame*, interestingly he was not responsible for the 1990s animated classic *Victor & Hugo: Bunglers in Crime*.

Previous birthday celebrations were undoubtedly spent with drinking buddy Quasimodo, who he once met in a bar in Paris. On that fabled day, Quasimodo ordered a blended whisky, and having looked through his limited selection of Scotch, the bartender said: 'Bell's OK?' Quasimodo replied: 'Mind your own f*cking business!'

Mark Victor's birthday with a Bell's Blended Scotch Whisky. Add some ice and top with soda water for a highball.

27 FEBRUARY

Henri IV crowned King of France (1594)

FRENCH 95 –

Apparently, the mummified skull of King Henri was found in a tax collector's attic in 2012. So now you know.

In 1594, Henri became the first French king from the House of Bourbon, named after the French spa town, Bourbon-l'Archambault in the Auvergne-Rhône-Alpes region. Interestingly, the town also gives its name to Bourbon County, Kentucky, named in honour of French King Louis XVI, who helped the Americans beat the Brits during the War of Independence – Louis, too, being a Bourbon.

And it is for this long-winded and tenuous reason that we suggest you have a French 95, a clever bourbon twist on the French 75 cocktail.

~

15ml (½ fl oz) bourbon | 15ml (½ fl oz) fresh lemon juice
15ml (½ fl oz) orange juice | 10ml (½ fl oz) sugar syrup
Ice | Champagne, to top up

- Shake the bourbon, lemon juice, orange juice and sugar syrup in a cocktail shaker with some ice. Strain into a champagne flute and top up with champagne.

28 FEBRUARY

Andalucía Day, Spain

ALHAMBRA 1925 –

With Protected Denomination of Origin on everything from *jamón serrano* and asparagus to olive oil and vinegar, the Andalucían locals pair gorgeous grub with fine wines, sherry and, increasingly, quality craft beer. Among them you'll find Alhambra, a brewer of clean, crisp Pilsner styles that make useful alternatives to a white wine if you're on the olives. But for something with a little more malt to match with meat, try Alhambra 1925. Using a controlled slow fermentation lasting more than 35 days, it weighs in at 6.4 per cent ABV and has the heft to honour to your *jamón*.

---

29 FEBRUARY

Leap Year Day

LEAP YEAR MARTINI –

While tackling the maths around a Leap Year, we got busy thumping buttons on a calculator and cross-referenced a Gregorian calendar and, frankly, it started to hurt our simple brains.

So, we've put the weird extra day in February down to magic.

If you're one of those magical people who have a birthday on a day that sometimes doesn't exist, you'll be pleased to learn that there is a cocktail created purely for this occasion. Harry Craddock was a pioneering bartender in the early 20th century. He wrote the bartender bible *The Savoy Cocktail Book* and invented the Leap Year Martini on this day in 1928 for the Savoy Hotel.

~

60ml (2 fl oz) London Dry Gin | 15ml (½ fl oz) Grand Marnier
15ml (½ fl oz) sweet vermouth | 10ml (½ fl oz) fresh lemon juice
Ice | Lemon zest twist, to garnish

- Shake the gin, Grand Marnier, vermouth and lemon juice in a cocktail shaker with some ice. Strain into a chilled cocktail glass. Garnish with a lemon zest twist and serve.

MARCH

1 MARCH

NATIONAL PIG DAY (AMERICA)

BENTON'S OLD FASHIONED -

Here are some pig facts you may not know:

Globally, we eat 400 billion tonnes of pork annually – a third more than beef. And a lot more than chicken, whose consumption figures are, quite frankly, poultry in comparison.

According to our favourite porcine-based 'novella', two out of three little pigs use inferior building materials to build their homes, yet, contrary to their reputation as dirty animals, they're the only farm animals who separate their living space into two: a latrine area and pristine sleeping quarters.

Why do pigs cover themselves in mud? So they don't get sunburn. That's true. They're also pretty useful after they're slaughtered, making everything from insulin and porcelain to anti-wrinkle cream (or should that be oinkment?) and bullets.

And bacon.

So here's a bacon-infused cocktail to mark the occasion.

~

40ml (1¼ fl oz) bacon fat-infused Bourbon (see below)
1 teaspoon maple syrup | Ice
2 dashes of Angostura bitters | Orange slice, to garnish

- Stir the bourbon and maple syrup in a mixing glass with ice. Add the bitters, then stir and strain into a chilled Old Fashioned glass filled with ice. Garnish with an orange slice.

To make the bacon-infused bourbon: Fry four strips of smoky bacon in a pan. Strain 30ml (1 fl oz) of the bacon fat into a container filled with 70cl (25 fl oz) bourbon. Leave to steep overnight at room temperature before placing the bottle in the freezer. Once the fat has congealed inside the bottle, strain the bourbon into a new, clean bottle and discard the fat.

---

2 MARCH

THE FIRST EVER CONCORDE FLIGHT (1969)

DOM PERIGNON (PREFERABLY 1969)
– AS SERVED ON CONCORDE –

A collaboration between French and British aeronautical industries, Concorde was a high-tech, supersonic work of genius. Capable of cruising at more than twice the speed of sound, it could fly from London to New York in three hours.

The thing is, few travellers could afford the extortionate tickets, and no other airlines were interested in a plane that not only guzzled 6,770 gallons of fuel every hour, but was also banned from huge swathes of airspace on account of its eardrum-shattering sonic boom.

After a tragic crash in 2000, Concorde was retired in 2003 as one of the worst civil investments in the history of humanity.

---

3 MARCH

BUFFALO BILL MEETS POPE LEO XIII (1890)

TATANKA –

Born in 1847, William Frederick Cody, aka Buffalo Bill, earned his nickname after he killed 4,282 buffalo between 1867 and 1868, supplying meat for the Kansas Pacific Railroad.

In 1883, he formed 'Buffalo Bill's Wild West', a circus spectacular featuring Native Americans and cowboys doing a whole lot of knife-throwing, sharp-shooting, rough-riding and that kind of thing.

Touring his show all over Europe and America, he visited the Vatican in 1890 and met Pope Leo XIII.

The moment was described in the *New York Herald*:

> *One of the strangest events ever to occur within the austere walls of the Vatican was the sensational entry made this morning by Buffalo Bill leading his cowboys and Indians.*
>
> *Among the immortal frescoes of Michelangelo and Raphael, and in the midst of the oldest Roman aristocracy, suddenly a band of savages appeared, their faces painted, covered with feathers and arms, armed with axes and knives.*

Commemorate the event with a Tatanka, a simple mix of apple juice and Żubrówka vodka (which contains a blade of bison grass in each bottle). The American bison and the buffalo are essentially the same animal: the only difference is, as you know, one can't wash one's hands in a buffalo.

---

4 MARCH

DEATH OF JOHN CANDY (1994)

ORANGE WHIP –

John Candy was an incredibly funny man. Canadian-born, six foot three, and sometimes weighing as much as twenty stone, he brought an anarchic affability to an array of iconic eighties classics, including *Planes, Trains & Automobiles* and *Uncle Buck*.

In *Blues Brothers*, he played Burton Mercer, a parole officer assisting the Illinois Police department in their pursuit of Jake and Elwood Blues. In a scene that has achieved comedy cult status, Candy orders a round of Orange Whips for his officers. 'Who wants an orange whip? Orange whip? Orange whip? Three orange whips.'

Why? Someone on set was the son of an Orange Whip employee and had asked director John Landis if it could possibly be mentioned in the film. Landis mentioned this to Candy, who squeezed it into the scene.

This is an alcoholic version.

~

50ml (1½ fl oz) single cream | 25ml (¾ fl oz) vodka
25ml (¾ fl oz) rum | 100ml (3½ fl oz) orange juice | Ice

- Blend all the ingredients (except the ice) in a blender for 30 seconds. Pour into a glass filled with ice and stir.

---

5 MARCH

TOBACCO ARRIVES IN EUROPE (1558)

OAXACA OLD FASHIONED -

In the 16th century, Spanish physician Francisco Fernandes returned to Europe from Latin America with tonnes of tobacco leaves.

Just as the Mayans had been doing for centuries, Europeans immediately embraced its medicinal properties. In London, they used it to resuscitate drowning victims – by blowing smoke up their arses.

Convinced warm tobacco fumes could combat cold and drowsiness, physicians performed smoke enemas using bellows inserted into the anus of sodden victims. Before long, tobacco-smoke enema kits were being provided along the River Thames by the Royal Humane Society. This suggests it must have worked. Either that, or it was simply something fun to do during a weekend stroll along the towpath.

Soon after, doctors realised that tobacco was actually very bad for you. And it still is, so please don't smoke – and certainly don't put it up your bottom.

Instead, try a Oaxaca Old Fashioned as created by Phil Ward in New York, using smoky tequila and even smokier Mexican Mezcal.

~

40ml (1¼ fl oz) reposado tequila | 12.5ml (½ fl oz) mezcal
5ml (¼ fl oz) agave nectar | 2 dashes of Angostura bitters
Ice | Orange zest twist, to garnish

- Pour the liquid ingredients into an Old Fashioned glass containing a large ice cube. Stir until chilled. Flame an orange twist over the surface of the drink.

---

6 MARCH

Day Of The Dude

WHITE RUSSIAN –

Directed by the Cohen Brothers and starring Jeff Bridges, the 1998 film *The Big Lebowski* inspired an unlikely cult following that morphed into a 'recognised' religion in 2005. On this, their most sacred day, more than 450,000 Dudeists indulge in the passions of Jeffrey 'The Dude' Lebowski, an unemployed weed-smoking slacker who shuffles through life in a dressing gown, playing ten-pin bowling and drinking White Russians.

~

60ml (2 fl oz) vodka | 30ml (1 fl oz) Káhlua coffee liqueur
1 tablespoon double cream | 1 tablespoon full-fat milk
Ice | Pinch of ground nutmeg

- The Dude simply poured the ingredients into a glass and stirred.

---

7 MARCH

Birth Of Sir Ranulph Fiennes (1944)

SIR RANULPH FIENNES' GREAT BRITISH RUM –

Tough-man toff Sir Ranulph Fiennes is arguably the world's greatest living explorer. He is the first person to walk across Antarctica unsupported and the oldest man to climb Everest. In

2003, he ran seven marathons in seven continents in seven days, just months after undergoing double bypass surgery following a heart attack. He was 59.

During a 2000 solo ramble to the North Pole, Fiennes fell through the ice and got severe frostbite on his hands. Forced to abandon his quest, he returned to Britain, where the pain became unbearable. So, Fiennes went to his shed and cut his frostbitten digits off with a hacksaw … and no anaesthetic.

Wish him Happy Birthday with 'two fingers' of his very own Great British Rum which, during distillation, adds wood varieties hailing from his most epic adventures: sequoia from Canada, Norwegian pine and palm from Oma. But not a piece of timber from his shed.

---

8 MARCH

INTERNATIONAL WOMEN'S DAY

HANKY PANKY –

This classic cocktail was created by Ada 'Coley' Coleman, who held the esteemed position of head bartender of the Savoy Hotel's iconic American Bar between 1903 and 1926, one of only two women to ever do so.

When the late Charles Hawtrey, a comedic actor, asked Coleman for a drink 'with a bit of punch in it' she gave him this. He sipped it, then, draining the glass, exclaimed, 'By Jove! That is the real hanky-panky!'

Renowned for her raucous wit, trash talking and deftly made drinks, 'Coley' was a consummate hostess who kept the likes of Charlie Chaplin, Marlene Dietrich and Mark Twain in their cups.

Yet, despite 23 years at The Savoy, Coleman was only credited with one recipe in the illustrious *Savoy Cocktail Book*, which was published just seven years after she left.

~

45ml gin (1½ fl oz) | 45ml (1½ fl oz) sweet vermouth
2 dashes of Fernet-Branca | Ice | Orange zest twist, to garnish

- Stir the gin, vermouth and Fernet-Branca with ice in a mixing glass, then strain into a martini glass and garnish with an orange zest twist.

---

9 MARCH

JAPANESE SOLDIER HIROO ONODA STOPS FIGHTING SECOND WORLD WAR (1974)

YAMAZAKI 1979 MIZUNARA OAK

– BOTTLED AFTER 29 YEARS IN OAK –

In early 1945, with Japan facing certain defeat in the Second World War, Emperor Hirohito dispatched troops to the small Philippine island of Lubang in a desperate last roll of the dice to try to curtail American advances.

It was a suicide mission, and the soldiers knew it. Most surrendered, others were killed, but Lieutenant Hiroo Onoda and three other soldiers were determined to stay and fight.

Retreating to the jungle and surviving on shrubs and bananas, they took pot-shots at American patrols, disrupted supply lines and, believing them to be enemies, assassinated innocent locals, also burning their crops and nicking their cows.

When, six months later, the war ended with the atomic destruction of Hiroshima and Nagasaki, no one told Onoda and his twig-nibbling chums, and they continued their guerrilla warfare.

In 1950, 'come home' leaflets were air-dropped into the jungle, but Onoda dismissed them as sneaky American propaganda. He also refused to believe letters from his family, the Philippine government and even a signed testimony from Hirohito himself.

By 1972, all of his companions had been killed, but Onoda kept on fighting a war that had ended 25 years ago, oblivious to the assassination of John F Kennedy, the building of the Berlin Wall, the Moon landing, and Queens Park Rangers winning the League Cup in 1967.

After several failed attempts to find him, Onoda was finally tracked down by an eccentric, self-styled 'adventurer-explorer' called Norio Suzuki. Suzuki was, by all accounts, a few sushi rolls short of a bento box, but after just four days wandering around the jungle repeatedly screaming Onoda's name, he discovered the long 'lost' lieutenant.

The unlikely duo stayed in the jungle for weeks, discussing all the world events that Onoda had missed – most notably QPR overturning a two-goal deficit and scoring three times in just 18 minutes to become the first Third Division team to win a Wembley final. That the final goal of a remarkable comeback was scored by a player called Lazarus simply blew Onoda's mind.

He finally believed the war was over when his commanding officer returned to Lubang and relieved him of his duties. Onoda, emaciated, his uniform in rags, saluted and broke down in tears.

Roaring crowds greeted Onoda, but sadly, he simply couldn't hack it in his modernised homeland. The nation he'd been fighting for had changed too much and, in 1980, he moved to Brazil, where he died aged 91 – not before returning to Lubang and giving $10,000 to a local school.

---

10 MARCH

Publication of The Godfather by Mario Puzo (1969)

THE GODFATHER COCKTAIL –

We're going to make you an offer you can't refuse: a robust digestif much loved by Marlon Brando, who played Don Corleone in the eponymous film.

~

60ml (2 fl oz) blended Scotch | 20ml (¾ fl oz) Disaronno Amaretto
Ice | Orange zest twist, to garnish

- Stir the Scotch and Amaretto in a mixing glass with ice, then strain into an ice-filled Old Fashioned glass. Finish with an orange twist.

11 MARCH

BIRTH OF DOUGLAS ADAMS (1952)

PAN GALACTIC GARGLE BLASTER –

According to Douglas Adams' cult comic novel *The Hitchhiker's Guide to the Galaxy*, never drink more than two of these 'unless you are a thirty-tonne mega elephant with bronchial pneumonia'.

~

10 fresh basil leaves | 20ml (¾ fl oz) vodka
20ml (¾ fl oz) blackberry liqueur
20ml (¾ fl oz) Grand Marnier liqueur
50ml (1½ fl oz) cranberry juice
50ml (1½ fl oz) strawberry purée | Ice

- Shake the ingredients in a cocktail shaker filled with ice, then strain into a large glass filled with crushed ice and serve.

12 MARCH

BIRTH OF JACK KEROUAC (1922)

BOILERMAKER –

Jack Kerouac is best known for writing *On The Road*, the greatest ever road-trip novel. This semi-autobiographical, hedonistic journey in search of individuality, freedom and some kind of meaningful happiness was bashed out in just three weeks on a single, long roll of paper and, crucially, loads of top-class amphetamines.

Kerouac catalysed his creativity in the dive bars and dingy cantinas of 1950s America and his beloved Mexico. His favourite drink was a Boilermaker, the ultimate blue-collar bar call, consisting of a beer chased with a whiskey.

A diagnosed schizophrenic, Kerouac's relationship with drink was often rocky. Two years before he died, he declared, 'I plan to drink myself to death.' Sure enough, in 1969, aged just 47, Kerouac's liver packed in – succinctly answering his own question: 'Why on earth aren't people continually drunk? I want ecstasy of the mind all the time.'

---

13 MARCH

Opening of the Waldorf Astoria, New York (1893)

ROB ROY –

At one stage the biggest hotel in the world, the Waldorf Astoria is the epicurean birthplace of the Waldorf Salad, Eggs Benedict and Thousand Island Dressing. Before being torn down to make way for the Empire State Building, it also created the classic Rob Roy cocktail (essentially a Manhattan made with Scotch) – named after the eponymous operetta that opened in New York in 1894.

~

45ml (1½ fl oz) Scotch whisky | 15ml (½ fl oz) sweet vermouth
Dash of Angostura bitters | Ice | Cherry, to garnish

- Stir all the ingredients except the cherry in a mixing glass and drain into a Martini glass or coupe. Garnish with the cherry and serve.

---

14 MARCH

National Pi Day

MIKKELLER BEER GEEK BREAKFAST OATMEAL STOUT –

Zip your anoraks right up to the top and hail the global celebration of Pi (Greek letter 'π').

Pi, it says here, is the mathematic symbol representing the ratio of the circumference of a circle to its diameter – approximately 3.14159. As every white-coated boffin will tell you, however, that's just the abbreviated version.

The full Pi figure continues infinitely, reaching more than 22 trillion digits beyond its decimal point – without repetition or pattern. On Pi Day, maths nerds like to recite as many of the infinite digits of Pi as they can. Crazy guys.

On 21 March 2015, an Indian chap called Rajveer Meena set a world record by reciting Pi to 70,000 decimal places wearing a blindfold. It took him 10 hours. You can watch it on YouTube if you really want. Alternatively, just take our word for it.

If any males who are truly passionate about Pi are finding it difficult to find a sexual partner (and one can't think why that would be the case), they can always splash on a bit of 'Pi', a cologne produced by Givenchy, which claims to enhance the 'attractiveness of intelligent and visionary men'.

---

15 MARCH

PERCY SHAW OPENS HIS CAT'S EYE FACTORY IN HALIFAX (1935)

WORTHINGTON WHITE SHIELD –

Percy Shaw, an inventor from Yorkshire, created cat's eyes, the reflective studs enabling drivers to follow the road in the dark.

In 1933, Shaw was driving home through thick fog when, about to veer into a ditch, the eyes of a cat sitting on a fence reflected his headlights, and, in doing so, illuminated the road in front of him.

Suitably inspired, Shaw designed an ingenious contraption that not only provided illumination but also cleaned itself every time a car drove over it. He christened these creations 'cat's eyes'.

Twenty million cat's eyes were laid in Britain during the Second World War, and the rather eccentric Shaw became a millionaire. He threw lavish parties, where, instead of champagne, he'd serve his favourite beer, Worthington White Shield, a classic India Pale Ale.

In California, the equivalent of cat's eyes are called 'Botts' dots' after their inventor Dr Elbert Botts. This reminds us of our favourite gag from British comedian Ken Dodd: 'The man who invented cat's eyes got the idea when he saw cat's eyes in his headlights. If the cat had been going the other way, he'd have invented the pencil sharpener.'

---

16 MARCH

CAMRA FORMED (1971)

PINT OF CASK ALE –

Cask ale is Britain's finest ever invention. It's better than penicillin, the Internet, the corkscrew, Marmite and Viagra put together – and we should know. Recently, we actually *did* put them all together and ended up in A&E. Again.

What is cask ale? Also known as 'real' ale, it's unfiltered and unpasteurised hand-pulled beer that undergoes secondary fermentation in the barrel. Properly brewed and lovingly looked after, cask ale can be brilliant. But it almost disappeared in the 1970s, chased out of pubs by crap keg ales and lacklustre lagers.

Disillusioned with the state of their favourite drink, a quartet of cask ale lovers – Michael Hardman, Graham Lees, Jim Makin and Bill Mellor – got together at Kruger's bar in Dunquin, Kerry, Ireland and formed the Campaign for the Revitalisation of Ale (CAMRA), later to be renamed the Campaign for Real Ale.

More than fifty years later, CAMRA is one of the biggest consumer organisations in the country and continues to preserve and champion good pubs and cask ales.

17 MARCH

St Patrick's Day

JAMESON REDBREAST WHISKEY –

In the fifth century, Ireland suffered from a reptile dysfunction – it happens to the best of us.

Pesky pagan snakes were everywhere, slippery anti-Christian evangelists making a nuisance of themselves, shedding their skin, swallowing hamsters whole, hypnotising Mowgli, sticking their tongues out at everyone. That kind of thing.

But then St Patrick came along. With a grinning pig under one arm, and waving a knobbly stick with the other, he chased the snakes into the sea, all while wearing gold-buckled shoes that turn up at the end, scaring the Bejaysus out of them with his novelty green felt hat and false red beard, etc. (and numerous other Irish clichés used by lazy writers padding out a St Patrick's Day drink entry).

There. That should do it.

18 MARCH

Official Opening of the London–Paris Telephone System (1891)

THE FRENCH CONNECTION –

Only in extremely rare (and generally brief) periods in history have relations between Britain and France been anything but guarded, if not outright aggressive.

The early 1890s were one such period. After centuries of fisticuffs, the two nations decided that they wanted to be buddies. So, a full fifteen years after Alexander Bell invented the telephone, they created a cross-Channel phone service using under-sea cables.

Only two people could make a call at a time, and they had to use special booths set up in the centres of London and Paris. Sounds shit. Still, at least they were talking.

~

50ml (1½ fl oz) Cognac | 25ml (¾ fl oz) Disaronno Amaretto | Ice

- Stir in a mixing glass with ice. Strain into a warm brandy snifter.

---

19 MARCH

BIRTH OF CAPTAIN RICHARD BURTON (1821)

MORNING GLORY FIZZ –

Captain Richard Burton was a posh Victorian polymath who somehow packed loads of lives into one: he was an adventurer, explorer, geographer, anthropologist, fencer, duellist, government assassin, soldier and spy who explored the world for 50 years.

Renowned as a prolific 'swordsman' with a fondness for both sexes, Burton's amorous adventures earned him the moniker of 'Dirty Dick', a racy reputation rubber stamped when he translated the *Kama Sutra*.

While working as an undercover agent in India, his report into male brothels in Karachi had to be shelved for being a little too detailed – and almost ended his diplomatic career.

He also wrote *A History of Farting* (while wearing a fez) and translated *The Perfumed Garden*, a 15th-century Arabic sex manual, bemoaning its lack of bestiality and homosexuality.

Proficient in 40 different dialects and a cunning linguist, it's no surprise he liked a fancy liqueur too, beginning each day with a Morning Glory, an aptly-titled cocktail containing absinthe and whisky.

~

45ml (1½ fl oz) whisky | 5 dashes of absinthe
30ml (1 fl oz) lemon juice | 10ml (½ fl oz) sugar syrup
15ml (½ fl oz) egg white | Ice | Soda water, to top up

- Shake all the ingredients (except the soda water) hard in a cocktail shaker. Strain into a tall glass and top up with the soda water.

---

20 MARCH

ALBERT EINSTEIN PUBLISHES HIS GENERAL THEORY OF RELATIVITY (1916)

ONE STEIN OF SPATEN OKTOBERFEST LAGER –

Albert Einstein liked a beer. As a student in Switzerland, he ran the famous Olympia Academy drinking club, where fellow boffins would drink beer and debate science and philosophy.

When he was just 17, young Einstein's first job was installing electricity in Germany's oldest beer tent. How much did he charge? The whole tent. Little neutron joke for you there.

Not content with the theory of relativity, he also invented an alcohol-powered beer fridge. And when Princeton Hospital pathologist Dr Thomas Stoltz Harvey stole Einstein's brain (after he died, obviously), he stored it in a beer cooler.

Despite his fondness for beer, Einstein enjoyed it responsibly, believing that 'alcohol spoiled the mind'. In fact, he restricted himself to just one large glass of German beer a day. Yes, that's right. He really was an 'Ein Stein' man.

---

21 MARCH

FIRST EVER TWEET (2006)

CONISTON BLUEBIRD ALE –

In 2006, Twitter co-founder Jack Dorsey wrote the first tweet: 'just setting up my twttr'. In honour of the iconic logo, grab a Coniston Bluebird, an awesome award-winning session ale named after

Donald Campbell's speedboat, which fatally crashed on Coniston Water back in 1967. We don't have enough characters to tell you more. #Sorry

22 MARCH

PIERRE-JOSEPH PELLETIER BORN (1788)

BYRHH ON ICE –

Pierre-Joseph Pelletier successfully extracted quinine from the bark of the cinchona tree, thus saving the world from malaria.

Used in hundreds of tonics, aromatised wines, bitters and spirits, quinine has played a huge role in cocktail history – and, without it, 19th-century European expansion into the sub-continent would have been impossible.

Winston Churchill famously declared that the humble gin and tonic saved 'more Englishmen's lives, and minds, than all the doctors in the Empire'.

Today, quinine can be found in bitters, herbal liqueurs, vermouth and quinquina – a style of aromatised wine. One of our favourites is Byrhh, an oak-aged mix of red wine, coffee, bitter orange, cocoa, mistelle and quinine from the South of France.

Sip it over ice with a slice of orange as an early evening aperitif – when those pesky mosquitoes are at their most potent.

23 MARCH

PUBLICATION OF *LES LIAISONS DANGEREUSES* BY PIERRE CHODERLOS DE LACLOS (1782)

ABSINTHE AND GINGER ALE –

Two hundred years after this chilling tale of amoral decadence and betrayal ruffled the feathered wigs of the French aristocracy, it was turned into a critically acclaimed film starring John Malkovich.

Now, no one likes a name dropper, but we once had dinner with a Malkovich. No, really, we did. It was at the Edinburgh Festival in 2011, the night before we opened our first ever show – which was staged in a converted freight container.

Over some exquisite Bordeaux and fine food, discreetly muttering behind our fluttering fans, Malkovich gave us some much-needed acting tips and spoke of his fascination with absinthe – a hedonistic spirit that, we rather cleverly pointed out, epitomises the debauched, libertine and amoral lifestyle depicted in the famous 18th-century French novel that inspired the film.

We all had a wonderful evening, we really did, and he promised to come and see our show.

We never saw him again.

24 MARCH

Death of Maurice Flitcroft (2007)

JACK NICKLAUS CABERNET SAUVIGNON –

Maurice Flitcroft is history's most hopeless yet heroic golfer.

A wild-eyed 46-year-old Cumbrian crane driver, Flitcroft secured a place in the 1976 qualifying rounds of the prestigious British Open Golf Championship by pretending to be a professional.

He wasn't even an amateur, having only taken up the sport two years earlier using some mail-order clubs and a pair of plastic shoes, and practising on nearby school playing fields.

On the British Open's first day of qualifying, Flitcroft got lost on the way to the course and arrived with no time to practise. Having never set foot on a proper course in his life, in front of 'fellow' professionals, and amid the hush of hundreds of onlookers, he got down on hands and knees, placed the ball on the tee, then gave it an acutely underwhelming whack.

'A real high-flying disappointment' was how he described his first 'professional' shot: 'It was not a total disaster. It could have gone straight up, come down and hit an official on the head, but it didn't.'

One hundred and nineteen shots later, he carded a round of 121 – which, at a staggering 49-over-par, was the worst ever round in tournament history. He made headlines the next morning, but the papers mistook him for a prankster rather than a glorious dreamer who genuinely thought he could compete next to Jack Nicklaus, his hero.

Keith Mackenzie, the self-important rulebook made flesh in charge of the Open, was furious with Flitcroft and banned him from ever entering the Open again. But Flitcroft, who grew to despise Mackenzie, hoodwinked his nemesis for the next 14 years by entering the tournament using absurd pseudonyms, such as Gene Paychecki, Gerald Hoppy and Count Manfred von Hofmannstal – often while sporting a comedy moustache.

---

25 MARCH

John Lennon and Yoko Ono's 'Bed-in for Peace' Protest, Amsterdam (1969)

RUTTE CLASSIC DRY GIN AND TONIC WITH PEAS –

When we were little, we did several memorable protests in our beds. Thankfully, none of them were as widely reported in the press as that staged by John Lennon and Yoko Ono in 1969.

Days after getting married, the superstar couple turned their 'honeymoon' into a sedentary stand against the Vietnam War. This involved lying in a bed at the Hilton Hotel in Amsterdam, talking about peace, singing, playing guitar, ordering room service and probably, given Ono's vegan diet, enduring a Dutch oven or two.

Give peas a chance by garnishing your Dutch G&T with the green spherical seeds.

26 MARCH

ALBERT SAUVANT CRASHES HIS CRASH-PROOF PLANE (1932)

BRANDY SMASH –

In 1932, French inventor Albert Sauvant claimed to the world that he had created the first ever crash-proof airplane.

To prove it, he climbed into the cockpit of his cutting-edge contraption and was pushed off a 24m (79 ft) cliff in Nice.

Sauvant's crash-proof plane cascaded clumsily down the cliff face, ricocheting off rocks, and smashing into smithereens. When a stunned Sauvant emerged from the wreckage, he gingerly waved to the crowd and Pathé film crew, proudly announcing that he'd suffered nothing worse than a few bruises. He later declared his experiment a huge success. Mark the occasion with a Brandy Smash.

~

50ml (1½ fl oz) cognac | 5 fresh mint leaves
5ml (¼ fl oz) sugar syrup | Ice

- Shake all the ingredients in a cocktail shaker filled with ice, then strain into a glass filled with a large block of ice and serve.

27 MARCH

BIRTH OF CHARLES TANQUERAY (1810)

TANQUERAY GIN AND TONIC –

In the 18th century, gin had dropped English society to its knees. At its nadir in 1723, every man, woman and child (yes, child) were drinking two pints of gin a week – and it wasn't nice gin, it was crazy shit moonshine and mixed with everything from urine to sulphuric acid.

But a century later, after loads of laws had been passed to restrict gin production, the spirit's reputation was restored by Charles Tanqueray, who opened his first distillery in London when he was just 20 years old.

Son of a clergyman, Charles shunned a life in the church to create 'London Dry', a new continuously distilled style of gin made with juniper, coriander seeds, angelica root and liquorice.

It was superior to the 'Old Tom' gins that had hitherto dominated the market, and made gin acceptable again, even making it a bit posh. The recipe still only uses these four ingredients, and it's really rather nice. Accompany it with some decent tonic, loads of ice and a slice of lime.

28 MARCH

Sean 'Puff Daddy' Combs announces he now wants to be known as P. Diddy (2001)

CIROC VODKA –

Ciroc Vodka, distilled from French grapes, is part-owned by hip-hop legend Sean 'Puff Daddy' Combs.

P. Diddy?

Only when he drinks a lot of it.

We wrote that 'funny' in 2014. The following year, it was voted the worst joke at that Edinburgh Fringe Festival.

29 MARCH

FINAL US TROOPS LEAVE VIETNAM (1973)

PABST BLUE RIBBON –

In 1967, John 'Chick' Donohue was in a Manhattan bar chatting to the bartender, George Lynch, bemoaning the anti-war protests sweeping the country.

They were fed-up with the stick the guys in Vietnam were getting back home – especially as a lot of the guys fighting out there were fellas they'd grown up with in the Irish-American neighbourhood of Inwood.

When Lynch suggested someone should bring those boys a beer, let them know that people back home hadn't forgotten them, Chick downed his beer, slid off his bar stool and headed back to Vietnam – embarking on the most remarkable beer run in history.

The former marine grabbed a US Navy holdall and filled it with cans of Pabst Blue Ribbon. Armed with a list of names and unit numbers of half-a-dozen local lads, he wangled himself a berth on the next warship heading out from New York to 'Nam.

Two months later, he docked in Qui Nhon, wearing just jeans, trainers and a shirt. As luck would have it, the first military company he stumbled across included one of the guys on his list.

Pretending to be his buddy's step-brother, he smooth-talked his way around security and cracked open a cold (ish) one with Tom Collins, his astonished pal from back home.

Over the next two months, Chick's legendary lager-fuelled pilgrimage took him across a war-ravaged Vietnam, blagging seats on choppers, mail planes, military convoys and ships as he ticked off names and delivered beers to friends fighting on the front line who were wondering whether they'd ever get home.

Having got caught up in the brutal Tet Offensive, Chick arrived home in March 1968, where he never had to buy a beer again.

30 MARCH

Death of the Queen Mother (2002)

QUEEN MUM COCKTAIL –

The Queen Mother was an elbow-bending omnivore found almost constantly in her cups.

Notoriously promiscuous in her drinking, she drank Chartreuse, had aides carry her beloved Beefeater Gin hidden in hat boxes and became the highest-spending private buyer of Veuve Clicquot. In the 1930s, she founded the 'Windsor Wets' Club, a high-society drinking circle whose motto was '*Aqua vitae non aqua pura*' ('Spirits, not water').

At The Savoy, head bartender Joe Gilmore created the Savoy Royale in her honour – champagne, peach and strawberries. But we're going for this eponymous potent pick-me-up, which she often enjoyed *before* lunch.

~

Ice | 25ml (¾ fl oz) Beefeater Gin | 50ml (1½ fl oz) Dubonnet
Orange or lemon zest twist to garnish (optional)

- Fill a chilled glass with ice, pour in the gin and Dubonnet and stir. Garnish with a twist of lemon or orange zest, if you wish.

31 MARCH

Red Rum wins the Grand National (1973)

HORSE'S NECK –

The 1973 Grand National was the first of three won by the legendary Red Rum, Britain's most famous racehorse.

Having trailed another horse, called Crisp, by 15 lengths, Red Rum staged a remarkable comeback to win, in record time, one of the most exciting races in Grand National history.

L'Escargot, another gastronomic favourite of the French, finished an appropriately distant third.

~

Ice | 50ml (1½ fl oz) bourbon | Dash of Angostura bitters
130ml (4½ fl oz) ginger ale | Lemon zest twist, to garnish

- Pour the bourbon over ice in a glass, then add the bitters and top with ginger ale. Garnish with a lemon zest twist.

APRIL

1 APRIL

April Fools' Day

FOOL'S GOLD COCKTAIL –

In the late 16th century, the French flipped from the Julian calendar to the Gregorian, shunting the New Year from 1 April to January, but a collection of incurable thickos were so confused, they celebrated the New Year on the old date.

They were rightly dubbed fools, or, more accurately, 'April Fish', and to prove the point, all the clever clogs who remembered secretly stuck paper fish to the fools' backs and shouted: '*Poisson d'Avril!*'

What japes.

But after taking time to mullet over, we felt fish deserved more respect – we mean, for cod's hake, fish aren't stupid, they're brill. Regardless, this is un-trout-edly neither the time nor plaice to reel off fishy stories as we skate close to pun overindulgence – it must be sole-destroying for you to read such pointless pollacks. Also, let's not encourage April Fools' behaviour, since it is mostly for (s)prats.

Here's a Fool's Gold cocktail instead.

~

60ml (2 fl oz) Basil Hayden's bourbon
20ml (¾ fl oz) fresh lemon juice | 15ml (½ fl oz) limoncello
10ml (½ fl oz) sugar syrup | Ice
Lemon slice and fresh sage, to garnish

- Shake all the ingredients (except the garnish) in a cocktail shaker with ice, then strain into a rocks glass. Garnish with a lemon slice and fresh sage.

2 APRIL

FIRST PANDA CROSSING OPENED (1962)

BAMBOO COCKTAIL –

Who doesn't love pedestrian crossing factoids? No one doesn't love them. No one. So get this: when new 'panda crossing' road markings appeared outside London's Waterloo Station in 1962, they adopted exactly the same colour scheme as a zebra crossing, but used triangles instead of rectangular stripes. Pretty crazy, right?

Despite much fanfare, these crossings were phased out reasonably quickly. We don't really know why. Perhaps pandas aren't as popular as zebras; perhaps the triangles confused pedestrians; or perhaps they were just shit. There could be lots of reasons: these issues are rarely black or white.

~

45ml (1½ fl oz) dry sherry | 45ml (1½ fl oz) dry vermouth
2 dashes of Angostura orange bitters | Ice
Lemon zest twist, to garnish

- Stir the sherry, vermouth and bitters in a mixing glass over ice, then strain into a cocktail glass and garnish with a lemon zest twist.

3 APRIL

DEATH OF GRAHAM GREENE (1991)

J&B RARE BLENDED SCOTCH –

The writer Graham Greene made it to the age of 86, having mixed drinking and writing for 60 years, while also surviving a stint in Sierra Leone working for MI6, enduring a battle with a bipolar condition and riding his luck through 'games' of Russian Roulette with his brother on Berkhamsted Common.

Whisky permeates the pages of his work, most notably in the

drinking game he describes in *Our Man in Havana*. Protagonist Jim Wormwold takes on Cuban copper Captain Segura in a game of draughts, but converts his collection of whisky miniatures into the pieces. Each time a player takes his opponent's piece, he's forced to drink it, creating a natural handicap by weakening his strategic senses.

We don't propose you attempt this game in his honour: there are 16 miniatures on each side, and even our remedial grasp of maths allows us to conclude that lots of little drinks makes a big one in your tummy. Instead, raise a small glass of J&B Rare Blended Scotch. It was Greene's whisky of choice, and combines 42 single malts with grain whisky.

---

4 APRIL

BIRTH OF MAYA ANGELOU (1928)

SHERRY –

The writing of Maya Angelou has been of epic significance, with her poetry, plays, essays and seven influential autobiographies often highlighting the struggle and injustices faced by African-Americans. She played a role in the Civil Rights movement and was bestowed a Presidential Medal of Freedom, along with many other artistic accolades. And she drank sherry while she wrote.

Certainly the sherry thing is least important on Maya's incredible list of life achievements, but give it a whirl, because there are plenty of expressions to test, from dry fino all the way to rich and warm Pedro Ximinez.

5 APRIL

POON LIM FOUND ALIVE (1943)

RUM FLOAT –

When a German U- boat sank the merchant ship SS *Benlomond* in 1942, Chinese steward Poon Lim was the sole survivor. But no one knew until 5 April 1943, because, having clambered on to a raft as the boat went down, Poon drifted alone at sea for 133 days.

While his story isn't the origin of 'out on a lim(b)', Poon's endurance certainly embodies the phrase.

He was eventually spotted and saved by fishermen off the coast of Brazil. He lived to 72, and still holds the record for longest lone survival at sea. In his honour, then: a rum float.

~

100ml (3½ fl oz) rum | 2 big scoops of vanilla ice cream
5 dashes of Angostura bitters | Coca-Cola, to top up

- Put the rum, bitters and ice cream into Collins glass and top with the Coca-Cola. Serve with a straw and spoon.

6 APRIL

ABBA WIN EUROVISION (1974)

ABSOLUT ELYX –

In 1974, Swedish pop group ABBA were propelled to global stardom with their Eurovision-winning 'Waterloo' – and, we can all agree, it remains an 'absolute banger' of a track. Absolut Elyx is not only Swedish, but also a banger of a vodka, made from a single estate winter wheat and crafted using a fancy copper still.

7 APRIL

VIOLET GIBSON SHOOTS MUSSOLINI (1926)

RED SPOT 15-YEAR OLD IRISH WHISKEY –

Violet Gibson was a Dubliner who attempted to assassinate Mussolini as he walked through Rome. Sadly, of the two shots Violet fired in 1926, only one was on target – and it merely grazed his nose. Mussolini subsequently wreaked havoc for nearly 20 years more, while Violet spent the rest of her life in a mental institution in Northampton.

Not a particularly cheery story, but charge your glass with a shot of Irish whiskey and mark the moment with a spirit to match the mark on Mussolini's nose: Red Spot 15-year-old single-pot-still malt. Produced at the Midleton Distillery in Ireland, three different whiskeys are aged separately then blended. One is rested in Oloroso sherry casks, one in bourbon casks, one initially aged in bourbon casks before being finished in, would you believe it, Sicilian Marsala wine casks.

You couldn't write this stuff – and yet we have.

8 APRIL

DAVID COPPERFIELD MAKES THE STATUE OF LIBERTY VANISH (1983)

RUM AND COKE WITH A FLAMING LIME –

David Copperfield made the Statue of Liberty disappear in 1983. No, really, it disappeared. It's still there now, of course – but it wasn't. Sadly, you won't find much footage of the illusion, because it was so flippin' magic.

Since David loves theatrics, how about adding a flaming lime to a rum and coke in his honour? It's the perfect distraction tactic for any cocktail you serve, looking particularly terrific atop a tiki drink.

To prepare, simply cut a lime in half, hollow it out, fill with some over proof rum and then light as it floats on the surface of the drink.

Playing with fire, much like David's magical movement of 93-metre-high (305-foot) statues, can be a dangerous, so don't spill that flaming liquid on your pants. Because then your pants would be on fire. Like a liar.

---

9 APRIL

FINNISH LANGUAGE DAY

FINLANDIA VODKA –

Here's some Finish for you – never accuse us of not putting in the hours:

Salmiakki Koskenkorvan lakritsipohjaisen liköörin maku on haastava, mutta Koskenkorvan kylästä peräisin oleva ohra-vodka, jossa on raikasta lähdevettä ja kevyttä keksiä, on erinomainen. Hyvin tehty Suomi.

(Blame Google Translate if this is nonsense.)

---

10 APRIL

WITHNAIL & I RELEASED (1987)

FINE WINE –

This cult masterpiece is a nailed-on classic for the oenophile. Ignore the highly irresponsible drinking game it spawned (one that encourages fans to match protagonist Withnail's infamous thirst by working through a pint of cider, two and a half shots of gin, six glasses of sherry, thirteen whiskies, four pints of ale, one shot of lighter fluid and nine and a half glasses of red wine…)

Instead, drink less but better, and celebrate with a glass of your best red – because, as they filmed, the *Withnail* cast and crew helped director Bruce Robinson sip through his personal collection of 200 fine wines. In today's auction houses, Bruce's bottles would fetch a pretty penny. They saw off Châteaux Margaux, Beychevelle and Pétrus vintages including 1945 (an epic year), 1947, 1953, 1959 and 1961.

---

11 APRIL

FUTURISTS CREATE THE *TECHNICAL MANIFESTO OF FUTURIST PAINTING* (1910)

AVANVERA COCKTAIL –

Futurism was a disruptive movement underpinned by the belief that art, in varied forms, could transform Italian society.

Futurists saw drinking as a display of nationalistic power, with wine regarded as the 'fuel of the Nation', absinthe made in secret, and Italian elixirs hitherto enjoyed neat used as mixers.

Proponents would talk, drink and fight in Milanese bars such as Caffè Del Centro and Caffe Savini, and, as they ploughed through the bottles, they created an unusual array of Futurist cocktails.

Since the Futurists believed in the random nature of process, there were no fixed recipes – which sounds cool, but is irritating if you're writing a book. So here's a close approximation of the Avanvera.

~

30ml (1 fl oz) Cocchi Vermouth de Torino
30ml (1 fl oz) Italian brandy
10ml (½ fl oz) Strega | Ice | 5 banana slices

- Pour the vermouth, brandy and Strega into a rocks glass over ice and stir. Garnish with five slices of banana and serve.

12 APRIL

RELEASE OF *PLANET OF THE APES* (1968)

MONKEY SHOULDER WHISKY –

'Get your stinking paws off me, you damned dirty ape!' So screamed firearms enthusiast Charlton Heston in 1968 as he grappled with hostile and hairy hominoids in cinematic classic *Planet of the Apes*. The film was based on Pierre Boulle's *La Planète des Singes* and, thanks to its expanding franchise, has grossed more than $2 billion worldwide. Top banana.

Less top banana (for the cast) was the lack of CGI, since 200 chimp-clad extras had to don sweat-inducing masks while performing in the scorching Arizona desert. Costumes were worn during breaks and all food was puréed, consumed via straws.

Had we overseen refreshments, we might've stirred some whisky into those liquified egg-salad sarnies – opting, obviously, for Monkey Shoulder.

The whisky is named in honour of distillery workers, whose over-stretched arms would hang low after a long day turning malt with heavy shovels – trials and tribulations akin to wearing monkey clobber in the desert. It's a blend of three malts from Speyside distilleries, is spicy and sweet and can be sipped neat, accompanied by complimentary nits picked from your work colleagues.

13 APRIL

CASINO ROYALE PUBLISHED (1953)

AMERICANO –

James Bond fans tend to associate the less-than-secret agent with his shaken-not-stirred Martini, but the first cocktail he orders in the *Casino Royale* book is actually an Americano. True, the Vesper

Martini (see 28 May) gets a spotlight moment in the casino caper, but during the book Bond also orders Veuve Clicquot champagne, brandy and chilled neat vodka, proving his thirst is quenched by a wider range of beverages.

In the second Bond novel, *Live and Let Die*, he polishes off Polish vodka Martinis, with plenty of gin Martinis thrown in, but Old Fashioneds steal the show. Meanwhile, *Moonraker* sees him spaced out on a flurry of vodka Martinis, wine, neat vodka, Dom Perignon, cognac, whisky, brandy and soda and Scotch.

By this point, it's becoming clear that rather than being eclectic in his tastes, the secret agent simply opts for whatever is close to hand, and during *Diamonds are Forever* you wonder if it'll be the bottle rather than a bullet that kills him off. Here, he downs Irish coffees, Old Fashioneds, Stingers and Back Velvets, alongside countless Martinis, champagne, bourbon and beer.

And while the fifth book, *From Russia with Love*, opens with a much-needed strong coffee from London's De Bry, you sense things are heading south in more ways than one when he departs for Istanbul. During a thirty-minute layover in Rome, Bond sees off two Americanos; on arrival in Athens, he's hitting the ouzo; and during his first meal in Istanbul, he manages two dry Martinis before a half bottle of Calvet claret. Then it's onto the raki, before pairing a doner kebab with a bottle of Kavaklıdere wine, followed by vodka and tonic, slivovitz and Chianti Brolio, all topped off with a double vodka Martini.

By 1961's *Thunderball*, Bond opens with a note of regret over one drink too many – an 11th whisky the night before – and we assume he'll slow down. But, no: he doesn't learn his lesson, and instead ploughs through gin Martinis, Chianti, double Old Fashioneds and a Stinger. By the end of that adventure, it's fair to wonder how the booze-soaked spy can even see, let alone beat up baddies.

And so it goes on, from Dom Perignon in *The Spy Who Loved Me* to Miller High Life in *On Her Majesty's Secret Service*, Bond is as relentless as he is indiscriminate, making him one of the most indulgent literary characters of all time. Granted, he's not one to emulate, but for a drink in his honour, go back to that very first order of an Americano, which, like Bond, is a genuine classic.

~

Ice | 45ml (1½ fl oz) Campari | 45ml (1½ fl oz) sweet vermouth
Soda water, to top up | Orange slice, to garnish

- Fill a rocks or highball with ice, then stir in the Campari and vermouth. Top with soda water and garnish with an orange slice.

---

14 APRIL

ASSASSINATION OF ABRAHAM LINCOLN (1865)

BULLEIT BOURBON –

Four score and seven years ago (plus a century or so), American President Abraham Lincoln was shot while he watched *Our American Cousin* at Ford's Theatre in Washington. Born in Kentucky, Lincoln's father worked for the Boone whiskey company, only a few hours' horse ride today from where Augustus Bulleit first made his own bourbon in 1830.

While it might seem insensitive to recommend a bourbon with the name 'Bulleit' on the anniversary of a man who was shot, the synergy was a little difficult to ignore. Besides, when Tom Bulleit revived his great-grandfather's creation, he re-introduced a fine spirit, one that is rich in vanilla after a spell in new American oak and beautifully balances the sweet corn in the mash bill with a decent dose of spicy rye.

15 APRIL

THE *TITANIC* SINKS (1912)

TALISKER 10-YEAR-OLD –

The *Titanic*'s maiden journey, like many voyage vacations, was a Trans-Atlantic tear-up, a booze cruise beyond compare – or at least, it was until it walloped into an iceberg.

Bottles of beer, wine and spirits had been stored aboard in their tens of thousands. Rather ironically, bar menus boasted 'iced draught Munich lager', along with Rob Roy, Robert Burns and Bronx cocktails, champagne and Bordeaux. Inventories listed everything from yellow chartreuse and fine sherry to 'dry dry' gin and 10-year-old Scotch.

And, like any great party, the *Titanic* had one dedicated drinker who refused to go down without a fight, who kept wetting his lips long after the party was over, and survived well into the morning. That man was Charles Joughin, the ship's chief baker.

As the ship began to sink, Joughin charged his glass with the 10-year-old whisky, and, brimming with Dutch courage, he calmly ensured women and children were given a safe berth on the limited lifeboats. Having refused a seat himself, he then set about directing his staff to carry bread to the lifeboats. Finally, as the ship began its final plunge into the icy depths, he climbed to the highest point of the bow and calmly stepped off into the water without getting his hair wet. Remarkably, Charles managed to survive, spending three hours in the ocean, and while alcohol won't protect you from hyperthermia, it enabled Joughin to remain calm in the midst of chaos.

For a 10-year-old single malt, then, we suggest Talisker. The distillery was founded in 1827 by two brothers, Hugh and Kenneth MacAskill, after they rowed from their island of Eigg to Skye, initially to start lamb farming. Avoiding icebergs, they made it safely to Skye, but, sensing the lambs were a bit of a gambol, opted for distilling instead. Today, Talisker is a major tourist draw on the stunning island of Skye, and its single malt is revered by whisky lovers everywhere.

16 APRIL

Tally's Electric Theatre Opens (1902)

SEVENTH ART –

A movie-themed cocktail to raise in honour of Thomas Lincoln Tally, who opened LA's first cinema in 1902. Created by Australian bartender Andrew Bennett.

~

Handful of popcorn, plus extra to serve
50ml (1½ fl oz) Bacardí Carta Blanca
10ml (½ fl oz) yellow chartreuse | 20ml (¾ fl oz) lemon juice
15ml (½ fl oz) sugar syrup | Pinch of salt | Ice

- Muddle the popcorn in a cocktail shaker, then add all liquids, a pinch of salt and some ice. Shake hard for 10 seconds, then strain into a chilled coupe. Serve with a ramekin of popcorn on the side.

17 APRIL

Progressive Beer Duty Introduced

BEER FROM A SMALL BREWER OF YOUR CHOICE –

Brace yourself for some beguiling pillow talk about 'Progressive Beer Duty'. Ooh, yeah. Fruity duty. Hot stuff …

As unsexy as it sounds, though, Progressive Beer Duty (or the Small Brewers' Relief, as it was christened in the UK), has had a colossal impact on craft beer. In 2002, the UK government announced anyone producing under 5,000 hectolitres of beer a year would pay 50 per cent of the standard duty rate, paving the way for smaller brewers to invest in people and product. The move spawned a stronger and more creative assembly of start-ups – so support a small brewer and pick up some beer from your nearest today.

18 APRIL

BIRTH OF MALCOLM MARSHALL (1958)

DOORLY'S RUM –

Malcolm Marshall was an outstanding Barbadian fast bowler who is often named as one of the greatest players of all time. His homeland offers up phenomenal rums to match his prodigious talent, and we suggest you check out R. L. Seale's Foursquare Distillery.

Like Marshall, Seale is regarded as a leader in his field, but don't be stumped by his exceptional range of rums. For those just getting their eye in, try Doorly's, a rum with great length that will bowl you over. It's also a wonderful opener and all-rounder, so it won't catch you out. There's a spiced expression, too, delivering some lively bounce and spicy tickle through the gate. Does that last one make sense? Probably not. But howzat for a pacey guide? And don't go for a hat-trick of bottles: you'll be in danger of a collapse, or at least an aggressive follow-through. That's it, we've run out. Time for some ball tampering. Cricket.

19 APRIL

INAUGURAL MISS WORLD PAGEANT (1951)

DIPLOMÁTICO RESERVA EXCLUSIVA RUM –

The Miss World contest is a wholly outdated, misogynist misrepresentation of women. Even so, Venezuela has won more than any other nation, so ignore the event but admire the magnificence of this country while sipping Diplomático Reserva Exclusiva. This stunning Venezuelan rum is rich and smooth with a lot more class than a beauty contest.

20 APRIL

Waldo Day

WALDOS' SPECIAL ALE, LAGUNITAS BREWING COMPANY –

Waldo Day is the stoner's 'national holiday', an entire 24 hours dedicated to the celebration of marijuana. It was started by five Californian high-school students in 1971, who hung out on a wall by a Louis Pasteur statue at 4.20pm, before heading off for a post-study toke, or 'Louis break'. These chaps became the 'Waldos', and such was their commitment to the cause that their '420' code spread, and was eventually converted to the date 4/20 (20 April) and cemented in lore as Weed Day.

In 1993, two big beer fans and disciples of the 420 code were sparking up a massive jay in the hills near Lagunitas and decided to start a brewery – and from that unshackled brainstorm came the Lagunitas brewery, now in Petaluma.

In 2011, forty years after that first 420 meeting, the folks at Lagunitas called up the Waldos crew, inviting them to brew a beer honouring the origins of 420. The Waldos agreed, and Waldos' Special Ale was born, described by Lagunitas as 'the dankest and happiest beer' they've ever created. At 11 per cent ABV, this imperial IPA is a big, herbal, bitter beer, and is possibly best paired with a some weed. We wouldn't know, we don't do drugs. Drugs are for mugs.

21 APRIL

Event: Rome Founded (753 BCE)

AFFOGATO –

While Rome wasn't built in a day, it was possibly founded on this day in 753 BCE by mythical twins Romulus and Remus.

Abandoned as babies, the twins survived by suckling on a wolf mother. While that sounds a bit odd, lapping at the *canis lupus*'s low-hangers was a crafty method of using an actual wolf to keep the figurative wolf from the door. Wolf boobs secrete colostrum, a watery milk that contains important antibodies – so now you know what to say if you ever get caught in the act. We're coming for you, Bear Grylls.

In an entirely relevant drinks connection, then, when in Rome, drink an affogato. This eternal post-dinner favourite comprises a scoop of ice cream in a cup, with hot espresso poured on top. The ice cream can represent frozen wolf's milk (it might as well), and we suggest you add 25ml (¾ fl oz) grappa before you top the cup with coffee.

---

22 APRIL

EARTH DAY

WARNER'S HONEYBEE GIN –

If you're starting to question your purchase of this book, make a solid start to Earth Day by chucking it in the recycling. Further attempts to save the world can come through sipping Warner's Honeybee Gin, which includes a dollop of honey collected from the distillery's own beehives. A percentage of bottle sales support bee projects, which is great because, as we all know, bees are good.

---

23 APRIL

ST GEORGE'S DAY

BEEFEATER GIN –

If you're English, charge your glass with something patriotic as we remember your proud patron, Saint George. Although, it's worth saying George didn't actually ever visit England – he was Turkish born to Greek parents, climbed the ranks of the Roman army before he slayed a dragon in Libya and was executed in Israel. The English simply opted for the third-century martyr because they liked the cut of his Christian jib.

Anyhoo, George is protector of soldiers, archers, cavalry, chivalry, farmers and field workers, and riders and saddlers. The saint also watches over those suffering leprosy, plague and syphilis – although perhaps gives the syphilis crowd a bit of privacy. So, what a patron saint of England he is. And Georgia – he's patron saint of Georgia, too. And Ethiopia, as it happens. And Catalonia. And Portugal. Hitler also took a liking to him – he had a 'Legion of St George' using British traitors as troops.

So, ahem, raise a jar of something jingoistic, like gin, a quintessentially English spirit. Although gin isn't really 'English' either. It was partly conceived in Italy as an eau de vie with medicinal properties, using distilling techniques from the Middle East, before the use of juniper in spirits was popularised in France and made a commercial success by the Dutch. And, of course, when the great distiller Desmond Payne makes his classic London dry Beefeater, he sources juniper from Italy, coriander seeds from Bulgaria, angelica root from Belgium, liquorice from China, orange peel from Seville …

Hmm. Happy St George's Day.

---

24 APRIL

SIGHTING OF HALLEY'S COMET (1066)

COMET COCKTAIL –

When Halley's Comet streaked across the sky back in 1066, everyone shat themselves and there was a real 'end of the world' vibe. But it was all OK in the end, so enjoy a Comet cocktail.

~

45ml (1½ fl oz) cognac | 25ml (¾ fl oz) Mandarine Napoléon
25ml (¾ fl oz) grapefruit juice | Dash of Angostura bitters | Ice

- Shake all the ingredients in a cocktail shaker with ice, then strain into a cocktail glass.

25 APRIL

PHILADELPHIA PHILLIES INTRODUCE MASCOT PHILLIE PHANATIC (1978)

CLOVER CLUB -

Twitchers in 1978 might've observed that when a huge, green, flightless creature, with a distinctive, funnel-cum-phallic snout and invasive extendable tongue ran on to the Philadelphia Phillies Major League Baseball field, it bore more resemblance to a muppet than native East Coast American *animalia aves*. It was a fair conclusion, since rather than choosing a relevant local creature, this new team mascot, Phillie Phanatic, was conceived by Bonnie Erickson and Wayde Harrison (Erickson being the designer of Muppets Miss Piggy and Statler and Waldorf).

Despite the incongruous nature of this beast, and his tendency to use his weird darting tongue to lick the players' balls and touch fans without permission, he has won over fans with hilarious heckles, happy high fives and haphazard *hora* dance moves. You, too, can celebrate his existence while drinking a Clover Club, a cocktail conceived in Philadelphia's Bellevue Stratford Hotel in the late 19th century. This beautiful update comes from New York bar legend Julie Reiner.

~

50ml (1½ fl oz) gin | 20ml (¾ fl oz) lemon juice
20ml (¾ fl oz) raspberry syrup
½ egg white | Ice | Raspberry, to garnish

- Shake the gin, lemon juice, syrup and egg white hard in a cocktail shaker. Add the ice and shake hard again. Strain into a coupe and garnish with a raspberry.

26 APRIL

SEVEN SAMURAI RELEASED (1954)

CASAMIGOS BLANCO TEQUILA –

Few films can claim such influence on cinema as 1954's Japanese classic *Seven Samurai*, directed by Akira Kurosawa. *The Magnificent 7*, a Wild West remake, is the most obvious benefactor; *Star Wars* creator George Lucas is a fan; as is Quentin Tarantino; and you can find tributes in films ranging from *The Matrix* and *Lord of the Rings* to *Mad Max: Fury Road*.

But one film that knocks them into the shade is John Landis's *The Three Amigos*, starring *Saturday Night Live* legends Chevy Chase, Martin Short and Steve Martin. In a modest twist on Kurosawa's plot, Landis's samurai are out-of-work silent-movie actors, but otherwise it is its equal as a cinematic masterpiece.

We suggest you watch *The Three Amigos* with some Casamigos Reposado, which nods to both cinema and friends. Launched by great pals George Clooney (see 18 January), Rande Gerber and Mike Meldman, their agave bulbs are slow roasted for 72 hours before the juice is distilled, producing a smooth spirit with a slightly sweet finish, meaning you can sip it neat without shuddering, like the amigos.

27 APRIL

KING'S DAY, NETHERLANDS

KETEL ONE VODKA –

While the Dutch monarchy enjoyed its first run out in 1568, no one really gave two shits about the birthday of the king or queen. But in 1885, there was a change of heart, and Princess Wilhelmina enjoyed the very first *Prinsessedag*. It was such a huge national success that ever since they've held an annual holiday to commemorate the sitting monarch's birthday.

These days it's a proper shindig: everyone dresses in highly a-peeling orange garms, and drinks high concentrations of exceptional alcoholic beverages. And they are exceptional, because the Dutch have serious skills when it comes to spirits.

During those early royal celebrations, many of the best distillates were coming out of Schiedam, a quaint city near Rotterdam. Although it covered a modest 20 square kilometres, at its height the town accommodated an incredible 400 malt mills, roasting houses and distilleries, complete with epic 33-metre-high windmills, five of which are still standing.

The Second World War dented Schiedam's distilling dominance, but today you can visit a brilliant Jenever Museum for a complete history of gin and take a tour of the Nolet Distillery, operational since 1691 and famed for its genever.

Along with the native spirit *genever*, Nolet also makes Ketel One. This luxury vodka was created in the 1980s by Carolus Nolet. He had visited San Francisco to suss out the American gin market, only to discover a dearth of decent vodkas. Bartenders echoed his desire for a vodka with character, so Nolet created his own using 'Distilleerketel #1', a traditional copper pot still, to add an extra complexity and bite to the spirit.

Drink some of it today – and put on some orange clothes already.

---

28 APRIL

BIRTH OF HARPER LEE (1926)

TEQUILA MOCKINGBIRD COCKTAIL –

The American wordsmith wrote the Pulitzer Prize-winning 1960 classic *To Kill a Mockingbird*, so we're confident she would appreciate this punny take on the novel's title. Just so witty, right, Harper?

~

60ml (2 fl oz) tequila | 15ml (½ fl oz) crème de menthe
15ml (½ fl oz) lime juice | 7.5ml (¼ fl oz) sugar syrup
Ice | Mint leaf, to garnish

- Shake all the ingredients (except the garnish) in a cocktail shaker with ice. Strain into a cocktail glass and garnish with a mint leaf.

---

29 APRIL

HUNTER S. THOMPSON ATTENDS THE KENTUCKY DERBY (1970)

MINT JULEP –

After experiencing one of the most famous horse races in the world, journalist Hunter S. Thompson wrote up the event for *Scanlan's Monthly* in an article entitled: 'The Kentucky Derby is Decadent and Depraved'. He described it thus: 'A fantastic scene – thousands of people fainting, crying, copulating, trampling each other and fighting with broken whiskey bottles.'

Mercifully, a dose of health and safety has since tamed the event, but the whiskey remains, with more than 100,000 Mint Juleps sold there each year. Enjoy one or two, but – in case it needs saying – don't fight with broken whiskey bottles.

~

Ice | 10 fresh mint leaves, plus a sprig to garnish
60ml (2 fl oz) bourbon whiskey | 15ml (½ fl oz) sugar syrup
2 dashes of Angostura bitters

- Shake all the ingredients (except the garnish) in a cocktail shaker with ice, then strain into a julep cup half-filled with crushed ice. Stir, then top with more ice and stir again. Garnish with mint.

30 APRIL

Bugs Bunny's Debut (1938)

WHAT'S UP DOC? COCKTAIL –

What is up, doc? Well, we will tell you: named after creator Ben 'Bugs' Hardaway, Bugs Bunny is actually a hare, hence his original role in *Porky's Hare Hunt* in 1938. But since this longer-eared creature enjoys the same taxonomic rank as rabbits, let's avoid splitting hares. That is all. Folks.

~

45ml carrot juice | 15ml The King's Ginger | 15ml lemon juice
5ml cinnamon syrup | Ice | Carrot, to garnish

• Shake all the ingredients (except for the garnish) in a cocktail shaker with ice. Strain into an Old Fashioned glass over ice, then garnish with, well, a carrot – but don't get any stray hares in there.

MAY

1 MAY

May Day

PINT OF REAL ALE –

May Day means we're Morris Dancing, England's version of New Zealand's Haka.

Morris dancers wear cricket whites, boater hats and bells on their legs. They wave white handkerchiefs about, hit each other with sticks, and jump and twirl around two clay tobacco pipes laid criss-crossed on the floor.

They come accompanied by accordion and fiddle-players hell-bent on making your ears bleed. By the time the Morris dancing has finished, everyone needs a drink. Preferably real ale, served in a dimpled pint mug.

2 MAY

Death of Oliver Reed (1999)

THE STINGER –

Oliver Reed's drink-drenched life was tragic, but seldom dull.

A chivalrous ruffian with an insatiable appetite for high jinks, he was an incredible actor who, at one stage, was among the highest-paid stars in Hollywood.

He was born in south-west London's leafy suburbs, an indirect descendent, or so Reed claimed, of 17th-century Russian ruler Peter the Great, a fellow high-class hooligan.

'My idea of a good time,' Reed once told a TV interviewer, 'is to get a few friends together and get as drunk as we possibly can'. As his list of friends included Richard Harris, Richard Burton, George Best, Alex Higgins and *The Who* drummer Keith Moon, Reed's was a dangerous hobby.

Having somehow survived decades of drunken escapades, Reed's death came during a rare period of relative abstinence. Before securing a role in Ridley Scott's *Gladiator* as a slave dealer, Reed promised Scott he'd confine his drinking to weekends.

Yet the handsome hellraiser spent his last hours drinking in a Valetta tavern and arm-wrestling Maltese sailors. He collapsed and died. One of the last lines he uttered on screen was: 'You sold me queer giraffes.'

After his funeral, there was a ten-day wake at his favourite pub in Cork, Ireland. Reed had kindly put £10,000 behind the bar – 'but only for those who are crying'.

One of Reed's favourite drinks was crème de menthe. So let's have a Stinger.

~

60ml (2 fl oz) cognac | 25ml (¾ fl oz) crème de menthe | Ice

- Mix the cognac and crème de menthe in a mixing glass with ice, then strain into an Old Fashioned glass or tumbler filled to the top with crushed ice.

---

3 MAY

Birth of James Brown (1933)

J&B RARE BLENDED SCOTCH WHISKY –

Get down. Get on up. And do your thing. And then take it to the bridge. Can you take it to the bridge? Good God.

4 MAY

STAR WARS DAY ('MAY THE 4TH BE WITH YOU')

WOOKEY JACK IPA, FIRESTONE WALKER BREWING COMPANY –

The only bar scene in *Star Wars: A New Hope* takes place at the Mos Eisley cantina. This is not 'first date' territory. Full of angry-looking extra-terrestrials, jobbing freight pilots and the dodgy-looking dregs of intergalactic society, the service is slack and the house band – Figrin D'an and the Modal Nodes – have only got one song. The cantina doesn't serve droids. Or peanuts. Heaven knows what the gents' looks like.

So, instead of that weird red stuff everyone's drinking, we recommend an unfiltered Wookey Jack IPA from California's Firestone Walker.

5 MAY

BERT TRAUTMANN BREAKS HIS NECK (1956)

CLOUDWATER HELLES –

In the 1956 FA Cup Final, Manchester City's German goalkeeper, Bert Trautmann, broke his neck while bravely throwing himself at the feet of Birmingham City striker Peter Murphy. Assuming it was a tweaked muscle, he played on, clutching his neck, pulling off several miraculous saves and guiding his team to a 3–1 win. Three days later, an X-ray revealed he'd dislocated five vertebrae and was lucky not to be paralysed.

Breaking his neck in an FA Cup Final is not the most remarkable part of Trautmann's quite extraordinary life. Born in Bremen, Germany, a young Trautmann was an archetypal Aryan – blonde, blue-eyed and tall – and sports-obsessed.

Swept up in the zeitgeist, he joined the German army aged just 17 and endured an unspeakable war. On the Eastern Front, he was blown up several times, spent three days under rubble and was captured by the Russians and the French.

Escaping both times, he was finally apprehended by the English after jumping over a fence and landing at the feet of two soldiers having lunch. 'Hello, Fritz,' they said. 'Fancy a cup of tea?'

Following incarceration at a Lancastrian POW camp, where he first played in goal, Trautmann refused German post-war repatriation, staying in England to play football for St. Helens. His goalkeeping prowess soon attracted professional interest and Manchester City signed him in 1949.

City's fans were furious, appalled by the idea of a former enemy wearing the club colours so soon after the war. There were death threats, hate mail and confrontations in the street. And that was just his own fans.

When City played Fulham, a raging Craven Cottage shook with chants of 'Heil Hitler', but after a stunning performance between the sticks, Trautmann was given a standing ovation by both sets of fans while Fulham's players formed a spontaneous guard of honour.

Over 15 years, he played more than 500 games for Manchester City, and went on to receive an OBE for his work improving Anglo-German relations. When giving him his award, the Queen said, 'Ah, Herr Trautmann. I remember you. Have you still got that pain in your neck?'

This neck-soothing German Helles-style lager comes from one of Manchester's leading craft brewers, dovetailing soft Mancunian water with hops and Pilsner malt from Germany. It's a wonderful drop – not that Trautmann did many of those.

6 MAY

TOUSSAINT L'OUVERTURE SWITCHES SIDES (1794)

ESPRESSO MARTINI –

Former slave and military mastermind François-Dominique Toussaint L'Ouverture freed the slaves of Saint Domingue, the nation now known as Haiti.

On this day, in 1794, the 'Black Napoleon', who was initially fighting for Spain, transferred more than 4,000 of his troops to the French side on the condition France freed his people. The switch secured victory for the French yet, some years later, L'Ouverture was betrayed when they returned slavery to Haiti.

In 1803, imprisoned in the Jura mountains, L'Ouverture starved to death – but not before warning his captors: 'In overthrowing me, you have cut down only the trunk of the tree of liberty; it will spring up again from the roots, for they are numerous and deep.'

He was right. Within a year, Haitian freedom-fighters had secured independence after giving the French army a serious kicking. While hailed as the heroic catalyst for Haitian emancipation, Toussaint's lesser-known legacy comes in liquid form – an eponymous liqueur made from Arabica coffee beans infused in three-year-old Caribbean rum. It's great in an Espresso Martini.

~

30ml (1 fl oz) vodka | 30ml (1 fl oz) Toussaint Coffee Liqueur
30ml (1 fl oz) espresso coffee | Ice | Coffee beans, to garnish

- Shake all the ingredients (except the garnish) in a cocktail shaker with ice. Fine-strain into a chilled Martini glass. Garnish with coffee beans.

7 MAY

New Orleans is Founded (1718)

HURRICANE COCKTAIL –

This terrific tiki-tastic rum-soaked classic originated at Pat O'Brien's Bar in New Orleans back in the 1940s.

During the Second World War, the only way for bars to get hold of sought-after Scotch and cognac was to buy a job lot including loads of scratchy Caribbean rum. So Pat O'Brien, desperate to rid his cellar of ropey rum, designed a drink that not only used up stock, but also disguised the rough taste.

He then served it to thirsty sailors in a Hurricane glass, a ludicrously large, billowy goblet named after the old-fashioned hurricane lamp.

~

45ml (1½ fl oz) white rum | 25ml (¾ fl oz) dark rum
25ml (¾ fl oz) pineapple juice | 15ml (½ fl oz) orange juice
15ml (½ fl oz) lime juice | 15ml (½ fl oz) passionfruit syrup
Ice | Orange slice and cherry, to garnish

- Shake all the ingredients (except the garnish) in a cocktail shaker with ice. Strain and garnish with an orange slice and a cherry.

8 MAY

VE Day

ANYTHING

In the Second World War, none of the 'loser' leaders drank. Hidake Tojo was a teetotal totalitarian, Benito Mussolini only drank milk, and Adolf didn't moisten his Hitler moustache with anything intoxicating (apart from champagne, shortly before he killed himself).

In contrast, the 'winners' were proper booze-hounds. Winston

Churchill's devotion to discerning drink is legendary (see 30 November), Josef Stalin was a voracious vodka-drinker, while Franklin D. Roosevelt made himself daily Martinis and ended Prohibition.

Celebrate alcohol overcoming evil by doing what everyone did in 1945 – drinking whatever you can get your hands on. (Responsibly.)

---

9 MAY

FIRST EDITION OF THE *ARKANSAS STATE PRESS* PUBLISHED (1941)

BOURBON DAISY –

Inspirational civil rights leader Daisy Bates bravely helped end school segregation in America.

Born in southern Arkansas, Bates's childhood was a traumatic one: her mother was murdered by three white men during an attempted rape, and Bates was abandoned by her father and raised by family friends in a shotgun shack.

After attending a Black-only school where she made do with battered, out-of-date text books passed down from white schools, she founded the *Arkansas State Press* in Little Rock with her journalist husband.

Dedicated to the town's African-American community and unwavering in its championing of civil rights and coverage of racially motivated violence, the paper's fiery editorials focused mainly on the desegregation of schools in Arkansas, where racial prejudice had sunk deep into the psyche.

When the Supreme Court ruled that segregated schools were unconstitutional in 1954, Bates took Black children to register at local white schools. When the school denied entry, as they always did, she reported it in her paper.

Opponents threw rocks through her window and placed a burning cross on her roof, but Bates was not deterred. In September 1957, aided by a National Guard Convoy sent by President Eisenhower, she successfully enrolled nine courageous Black children into Little Rock Central High School, despite the threatening presence of a seething, snarling white mob outside.

'My eyes none too dry,' she wrote in her autobiography, 'I saw the parents with tears of happiness in their eyes as they watched the group drive off.'

The 'Little Rock Nine' marked a significant civil rights victory and forced the American government to fully implement the 1954 Supreme Court ruling. Alas, bigotry brought Daisy's business to its knees – with businesses unwilling to advertise, the Arkansas State Press closed in 1959.

Several of the Little Rock Nine went on to enjoy distinguished careers. In 1999, President Clinton awarded each of them the Congressional Gold Medal, while all of them were invited to President Barack Obama's 2009 inauguration. Raise a Bourbon Daisy cocktail in memory of Daisy Bates.

~

50ml (1½ fl oz) bourbon | 15ml (½ fl oz) triple sec
15ml (½ fl oz) fresh lemon juice | 7.5ml (¼ fl oz) grenadine syrup
Ice | Orange zest twist, to garnish

- Shake all the ingredients (except the garnish) in a cocktail shaker with ice. Strain into a Martini glass. Garnish with an orange twist.

---

10 MAY

INAUGURATION OF NELSON MANDELA (1994)

GRAHAM BECK BRUT NV –

More than 30 years after he was sentenced to life in jail, Nelson Mandela was sworn in as the first Black president of South Africa – after sweeping to power in the nation's first-ever multiracial parliamentary elections.

Attended by more than 4,000 guests, including high-profile dignitaries from across the political spectrum, Mandela's inauguration was televised to billions worldwide. He delivered a speech that, given the preceding decades of apartheid and violence, was remarkably free of personal bitterness.

With courage, compassion and an extraordinary generosity of spirit, Mandela declared: 'The time for the healing of the wounds has come. We shall build a society in which all South Africans, both Black and white, will be able to walk tall, without any fear in their hearts, assured of their inalienable right to human dignity – a rainbow nation at peace with itself and the world.'

During one of history's most significant speeches, the likes of Fidel Castro, Prince Philip and Boutros Boutros-Ghali were, one imagines, discreetly trying to grab the attention of waiting staff carrying trays furnished with flutes of fancy South African sparkling wine.

Graham Beck Brut NV, made from Chardonnay and Pinot Noir in the Breede River Valley, not far from Cape Town, was not only served on this historical day, but also at Barack Obama's presidential party – earning it the nickname of 'the President's Choice'.

---

11 MAY

BIRTH OF JACK MCAULIFFE (1945)

WEST COAST IPA –

America's craft-brewing revolution was quietly kickstarted by an unassuming man named Jack McAuliffe. Inspired by ales he'd enjoyed as a sailor stationed in Scotland, McAuliffe created America's first modern craft brewery in Sonoma, northern California, back in 1976. The New Albion Brewing Company's brew kit was welded and hammered together by hand and, using recipes McAuliffe had honed as a home-brewer, it made bottle-conditioned beers of British character.

Thing is, McAuliffe found himself a little too far ahead of the curve. With no money, no craft-brewing cohorts, and surrounded by a sea of light lager, he'd set sail in a one-man sieve. After six remarkable years, the business went bust.

But McAuliffe's innovation rolled out the red carpet for those that followed, and every beer in the craft movement is a liquid legacy of his pioneering efforts. Please salute his sacrifice with a pale ale.

12 MAY

International Limerick Day

A GLASS OF CIDER (SAUSAGE OPTIONAL) –

There was a young lady called Ida
When hungry, it was hard to abide her
But her mood got much better
When circumstance let her
Enjoy a big sausage in cider
– *Written by The Thinking Drinkers*

13 MAY

World Cocktail Day

SAZERAC –

While a topic of mass debate among hooch historians, many reckon the word 'cock-tail' was first defined on this day in 1806, in a New York publication called *The Balance, and Columbian Repository*.

It was described thus: 'Cocktail is a stimulating liquor, composed of spirits of any kind, sugar, water and bitters. It renders the heart stout and bold, at the same time that it fuddles the head.'

~

Absinthe, to coat | 60ml (2 fl oz) cognac
2.5ml sugar syrup | 2 dashes of Peychaud's bitters
Ice | Lemon zest twist, to garnish

- Coat the inside of a rocks glass with a little absinthe, then discard.
- Build the cognac, syrup and bitters in a mixing glass over ice, stirring slowly. Strain into the absinthe-rinsed glass. Squeeze the oil from the lemon zest twist over the drink's surface, then either drop it in or discard it.

---

14 MAY

Death of Gordon Bennett (1918)

THE BENNETT COCKTAIL –

Gordon Bennett, the inspiration behind the quintessentially English exclamation of incredulity, was the heir to the founder of the *New York Herald*.

As editor, Bennett famously commissioned Henry Morton Stanley to find Dr David Livingstone and return him from the African wilderness. Yet he's better remembered as a pompous, profligate playboy who would ostentatiously set money on fire, complaining it felt uncomfortable in his pocket.

Until recently, he also featured in the *Guinness Book of World Records* for 'the worst faux pas ever committed'. Highly inebriated, Bennett turned up at a party of New York socialites hosted by his fiancée and urinated in the fireplace, believing it was the toilet.

Bennett's fiancée immediately broke off the engagement, before her brother defeated Bennett in a duel. Riddled with shame, he moved to France – where being rude is perfectly acceptable.

~

50ml (1½ fl oz) gin | 20ml (¾ fl oz) lime juice
15ml (½ fl oz) sugar syrup | 2 dashes of Angostura bitters
Ice | Slice of lime, to garnish

- Shake all the ingredients (except the garnish) in a cocktail shaker with ice, then strain into a chilled coupe. Garnish with a lime slice.

15 MAY

LAS VEGAS FOUNDED (1905)

ATOMIC COCKTAIL –

Although it was founded in 1905, the 1950s were the boom years for Las Vegas – quite literally.

While Sinatra sipped Martinis with the Mob in salubrious hotels on the Strip, the US Atomic Energy Commission began exploding atomic bombs just 60 miles up the road in the Nevada desert.

At the time, the horrors of Hiroshima were still raw, and America was genuinely fearful of radiation. But this was Vegas, and the city's notorious Chamber of Commerce saw money in them thar mushroom clouds.

It printed calendars advertising detonation times, and the best venues with north-facing panoramic views from which to watch them. Casinos promoted parties where, despite genuine risk of radiation, revellers danced as explosions turned night into day.

The iconic Sands Hotel & Casino even staged a 'Miss Atomic Energy' beauty pageant, where contestants dressed up as mushroom clouds, while the Desert Inn, offering amazing views, created the suitably potent 'Atomic Cocktail'.

This continued until 1963, when above-ground nuclear trials ended. A shame, really, as we'd rather melt our own faces off than have to sit through a Celine Dion concert at Caesar's Palace again.

~

40ml (1¼ fl oz) vodka | 40ml (1¼ fl oz) cognac
20ml (¾ fl oz) Amontillado sherry | Ice
Champagne, to top up | Orange slice, to garnish

- Shake the vodka, cognac and sherry in a cocktail shaker with ice. Strain into a chilled Martini glass and top up with champagne, then garnish with an orange slice.

16 MAY

Death of Eliot Ness (1957)

CUTTY SARK PROHIBITION EDITION
BLENDED SCOTCH WHISKY –

Eliot Ness was the man who brought down Al 'Scarface' Capone during Prohibition.

Chicago-born in 1902, Ness headed up a group of Prohibition Bureau agents dubbed 'the Untouchables' (on account of their refusal to take bribes) and regularly raided Capone's illegal breweries and distilleries, costing him nearly $9 million.

Capone, in return, wasn't very nice to Ness. The 29-year-old agent was physically intimidated and had his car stolen several times. His office phone line was continuously tapped, and one of his informants was shot four times in the face.

Unwavering in his pursuit, however, Ness gathered enough evidence to charge Capone for both tax evasion and producing alcohol. But judges only chose to pursue the tax evasion indictments, as they didn't want jurors sympathising with someone who was giving everyone what they wanted.

Ness didn't like Prohibition. After busting Capone's bootlegging operations, he gave newspaper reporters confiscated stashes of booze in return for positive press – and he was a big fan of Cutty Sark whisky.

17 MAY

Wizard of Oz Published (1900)

BLUE NUN –

Judy Garland, star of the 1939 cinematic fairy tale based on L. Frank Baum's book, was a big fan of Blue Nun, the sweet German wine, long before it became 'cool' in the 80s. She often chased it with Ritalin, a nervous system stimulant commonly used to treat ADHD. We suggest you stick to the Blue Nun.

---

18 MAY

Napoleon Bonaparte becomes French Emperor (1804)

NAPOLEON MARGARITA –

When doctors delivered Napoleon Bonaparte, they discovered several teeth already implanted in his gums.

This rare condition, according to English folklore, is unique among babies destined to conquer the world. Which is exactly what the diminutive dictator tried to do.

Brilliant yet brutal, Napoleon conquered an empire stretching from Moscow to Portugal, and introduced liberal Enlightenment across Europe. He was also responsible for the callous deaths of millions.

Many attribute Napoleon's tyrannical behaviour to his frustration at his short stature but, at five foot seven, he was of average height for the time. The origins of the 'Napoleon Complex' instead derive from a British cartoonist called Gillray, who lampooned the French Emperor by drawing George III holding a tiny Napoleon in his hand.

While Napoleon didn't drink much, he inspired his friend and amateur distiller Antoine-François de Fourcroy to create Mandarine Napoléon, a cracking combination of aged cognacs and mandarin oranges from Napoleon's native Corsica.

While you can have it short, it makes a marvellous margarita.

~

30ml (1 fl oz) Mandarine Napoléon | 60ml (2 fl oz) tequila
30ml (1 fl oz) lime juice | 15ml (½ fl oz) agave syrup | Ice
Salt, for the glass, and lime zest twist, to garnish

- Shake all the ingredients (except the garnish) in a cocktail shaker with ice. Strain into a salt-rimmed rocks glass filled with ice cubes. Garnish with a lime twist.

---

19 MAY

Birth of André the Giant (1946)

BUDWEISER BUDVAR RESERVE, 'A STRONGER VERSION OF THE ORIGINAL BUDWEISER' –

When André René Rusimoff was 12 years old, he was six foot three, weighed over 240 pounds and didn't fit in his dad's car. He had to be taken to school in a truck driven by Samuel Beckett, who talked to him about cricket.

It sounds absurd, even by Beckettian standards, but these are the kind of big, bizarre things that happened in André's life. Born in the shadow of the French Alps, he was diagnosed with acromegaly, a rare glandular syndrome that accelerates growth – especially in one's head, hands and feet.

André was told he wouldn't live beyond his mid-twenties, but rather than sitting on the end of his long bed, staring mournfully at his oversized shoes, he bounced off the ropes of anguish and attacked life in an enormous unitard.

Around seven and a half feet tall and weighing 500 pounds, André the Giant dominated world wrestling for the best part of twenty years, became a Hollywood icon and famously set a world record for drinking the most beer in one sitting, seeing off 119 bottles of Budweiser in six hours.

André drank beer like mere mortals drink water, a feat made easier by the fact that American beer in the 80s actually tasted like water. His considerable size allowed him to consume, on average, 7,000 calories' worth of alcohol a day – the equivalent of 53 bottles of beer. Every. Single. Day.

20 MAY

Birth of Busta Rhymes (1972)

COURVOISIER ON THE ROCKS –

The French region of Cognac isn't somewhere you'd associate with gangster rap. Local grape growers may use 'hoes' to turn over the soil in their vineyards; perhaps there's a little bit of gun-toting during a duck shoot; 'bitches' (female dogs) are certainly great at snuffling out truffles during autumn … and, obviously, too much foie gras may well leave you needing to buy a pair of Bigger Smalls.

But that's where the association ends in these 'Endz'. Yet, without the American hip-hop scene, cognac would be in big trouble. In the 1990s and 2000s, African-American culture pretty much single-handedly saved the region from a devastating downturn in sales – with the likes of Jay-Z rhyming about the 'Yak'– most notably in 'Can't Knock the Hustle'.

The real boost, however, came in 2001, when Busta Rhymes (whose real name is Trevor George Smith) released 'Pass The Courvoisier II' in collaboration with fellow hip-hop legends P. Diddy and Pharrell.

Featuring lyrics far too crude to repeat here, the chart-topping hit extolled the qualities of the 300-year-old French distilled wine and sent sales of cognac, and Courvoisier in particular, soaring. Despite some very rude words and unwholesome associations with violence, misogyny, drugs and gun crime, hip hop's helping hand has been quietly welcomed by the major cognac houses.

Folk back in this rustic region of France, however, remain bemused by their brandy's musical association. With half of the population aged 50 years or more, any mention of hip hop is more likely to refer to a medical procedure in the pelvic area than N.W.A., P. Diddy or the Wu Tang Clan.

21 MAY

Birth of Mr. T (1952)

JIBBER JABBER WEST COAST PALE ALE, LONDON BEER FACTORY –

In his principal role as B. A. (Bad Attitude) Baracus in hit 1980s TV series *The A-Team*, Mr. T famously played a mechanic capable of building various military vehicles out of some dental floss and toilet roll. He also said things like 'I pity the fool' and 'Quit your jibber-jabber'.

22 MAY

International Paloma Day

PALOMA –

The Paloma is a terrific tequila-driven drink created by Don Javier Delgado Corona, the renowned owner of La Capilla, a legendary bar in the town of Tequila, Mexico. The drink is refreshingly unpretentious, easy to make and a much better way to taste tequila than downing it in one.

~

Salt, for the glass | Ice | 60ml (2 fl oz) tequila
15ml (½ fl oz) lime juice | Grapefruit soda, to top up
Grapefruit wedge, to garnish

- Rim a tall glass with salt, add ice cubes, tequila and lime juice in that order. Stir and top with grapefruit soda. Garnish with the grapefruit wedge.

23 MAY

National Turtle Day

TURTLE RECALL PALE ALE, BEDLAM BREWERY –

A guy goes into a bookshop and asks the lady behind the counter for a book about turtles.

'Hardback?' she says.

'Yeah,' he replies. 'And they've also got little heads.'

Have a can of Turtle Recall Pale Ale from the Bedlam Brewery in East Sussex – 10p per can helps save sea turtles.

24 MAY

National Escargot Day (US)

TROUBLETTE, CARACOLE BREWERY –

Humans have scoffed snails for more than 30,000 years, but no other nation gets behind gastropod-inspired gastronomy quite like France, where eating weird stuff is, along with shrugging, a national sport.

In French restaurants, escargot are served soaked in garlic, parsley, butter and wine, because soaking anything in garlic, parsley, butter and wine makes it tastes nice. Even snails.

High in protein, low in fat, snails qualify as a sustainable superfood, alongside goji berries and chia seeds. But the way they're cooked and killed has prevented the likes of Gwyneth Paltrow putting them on their Instagram accounts.

There are several ways of murdering snails. Some bury the snails in a bowl of salt and bid them farewell amid a seething fizzy froth. A more humane approach, and one that better preserves the flavour, involves gently suffocating them in an airtight jar placed in a fridge.

(We know it's not nice but, come on, they're only snails. And this is payback for eating all our plants back in 2006. So, try to ignore their forlorn-looking antennae as you close the fridge door.)

How to cook them? Well, snails are not, as you'd expect, fast food. You need to boil them for at least two and a half hours before frying them or baking them. Marinating them takes another few hours on top of that.

Eaten 'fresh', they can be meaty in texture yet delicate in flavour, and when slathered in garlic and parsley, they demand a crisp dry, minerally white, such as a Chablis. Or, if you really want to keep the snail theme going, pair with a bracing Belgian *witbier* called Troublette from the Caracole (Snail) Brewery, in Wallonia. It boasts a picture of a snail on the label, herbal notes, enough bitterness to cut through the butter and enough lively carbonation to lift the little fellows of the palate.

---

25 MAY

BIRTH OF COUNT CAMILLO NEGRONI (1868)

NEGRONI –

Count Camillo Negroni is thought to be the Italian behind this iconic aperitivo.

Having moved to New York in the 1890s, working as a rodeo cowboy and fencing tutor, Prohibition propelled him back to Florence, where he became a regular at Café Casoni.

Here, on one evening in 1919, he asked for something stronger than his customary Americano, and the bartender, Fosco Scarselli, replaced the soda with a touch of gin. To enhance the botanicals, Fosco replaced the lemon with orange, and the Negroni was born.

~

30ml (1 fl oz) gin | 30ml (1 fl oz) Campari
30ml (1 fl oz) sweet vermouth | Ice | Orange zest twist, to garnish

- Stir all ingredients (except the garnish) in a glass with ice. Strain into an ice-filled Old Fashioned glass. Garnish with an orange twist.

26 MAY

National Paper Plane Day

ORIGAMI SAKE –

Making a paper plane is officially origami. Celebrate with sake.

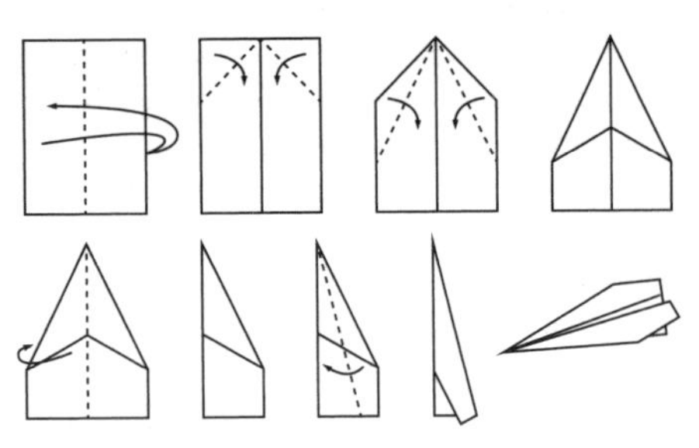

27 MAY

Birth of Wild Bill Hickock (1837)

AMERICAN WHISKEY. SHOT. STRAIGHTY –

Sharp-shooting Wild Bill Hickock, legendary Wild West folk hero, drank so much whiskey he earned the name Wild Bill Hiccup.

Despite this, the 'King of the Pistoleers' could reputedly hit an ace of spades at fifty paces. Others claim he was a sodden, sneaky gunslinger who simply shot his foes down when they weren't looking. Not really cowboy cricket.

He was killed in 1876 by a bullet in the back from Jack McCall as he played poker in Deadwood, South Dakota. Hickock was 39, but his hand was a winner: five-card draw, holding a pair of aces and a pair of eights. This saw him posthumously inducted into the Poker Hall of Fame.

28 MAY

Birth of Ian Fleming (1908)

VESPER MARTINI –

Few literary figures have done more for discerning drinking than James Bond creator, Ian Fleming.

A former secret service agent who'd worked in Naval intelligence during the Second World War, Fleming shared Bond's insatiable appetite for uncomplicated womanising, golf, gambling and fast cars – although his most spectacular stunt was reversing into a milk float.

While 007 was a ludicrously promiscuous drinker (yet somehow suffered just two hangovers in 14 novels) Fleming was famously overly fond of gin and, at his peak, consumed a bottle a day. When his doctor advised him to stop, he switched to bourbon.

In *Casino Royale*, written in 1953 from his Jamaican estate Goldeneye, Fleming created the Vesper Martini, reputedly introduced to him at Duke's Hotel in London. Lillet Blanc is a French aperitif wine, and *aperitif* is the French word for 'dentures'.

~

45ml (1½ fl oz) vodka | 15ml (½ fl oz) gin
5ml (¼ fl oz) Lillet Blanc | Ice | Orange peel twist, to garnish

• Shake all the ingredients (except the garnish) in a cocktail shaker with ice. Strain into a Martini glass and garnish with orange peel.

29 MAY

National Biscuit Day

GARIBALDI –

Giuseppe Garibaldi, the revolutionary instrumental in the unification of Italy, must surely be the only historical figure immortalised in both biscuit and cocktail form.

The classic biscuit treat consists of squashed currants sandwiched between two thin layers of biscuit dough, dotted with more squashed currants. The two cocktail ingredients, meanwhile, signify the unification of north and south Italy; Campari from Lombardy, representing the red shirts of Garibaldi's freedom fighters, and the juice of bright oranges sourced from southern Sicily.

The biscuit doesn't complement the cocktail. We tried it, and it really didn't work. You're better off with a nice cup of tea.

~

50ml (1½ fl oz) Campari | Ice
Freshly squeezed orange juice, to top up
Orange wedge, to garnish

- Pour the Campari into a small highball glass filled with ice. Add the OJ and stir, then garnish with an orange wedge.

---

30 MAY

DEATH OF VOLTAIRE (1778)

FLYING FRENCHMAN –

Voltaire, the famous French Enlightenment clever clogs, drank between 50 and 72 coffees a day.

If we were to consume this amount of caffeine, you'd find us in the foetal position, rocking back and forth, muttering to ourselves, with snot bubbles protruding from our noses. Or we'd be on the loo.

For Voltaire, however, a bountiful supply of the black stuff was crucial creative fuel for his prolific output as a philosopher, poet, dramatist, playwright, author and satirical scallywag who ruffled the feathered wigs of French society.

Many 'free speech' advocates wrongly credit Voltaire with the saying: 'I disagree with what you say, but I will defend to the death your right to say it.' This was, in fact, written by his biographer.

One thing he definitely did write, however, was a sardonic poem alleging the Duc d'Orléans of having hanky-panky with his own daughter. French 'free speech' at the time sadly didn't stretch to accusations of incest – especially if it involved the Regent of the Kingdom. Hence Voltaire's eleven-month stay in a windowless cell in the Bastille – where they certainly didn't serve skinny lattés.

~

30ml (1 fl oz) absinthe | 30ml (1 fl oz) coffee liqueur
30ml (1 fl oz) espresso | Ice

- Shake all the ingredients in a cocktail shaker with ice, then fine-strain into a Martini glass.

---

31 MAY

BIRTH OF CLINT EASTWOOD (1930)

GOLDEN CADILLAC –

In order to make Clint's day, we're having some ice cream.

In the 1980s, when Eastwood was Mayor of Carmel-by-the-Sea, California, he repealed a strange local law prohibiting the sale and consumption of ice cream on the town's boardwalk.

He did a bit of acting and directing in his time, too, but in honour of his services to small-town ice-cream entrepreneurs, we're getting our laughing gear round a Golden Cadillac made with ice cream instead of double cream – a cocktail created at Poor Red's Saloon in California.

~

25ml (¾ fl oz) Galliano | 25ml (¾ fl oz) white crème de cacao
35ml (1 fl oz) vanilla ice cream | Ice | Ground cinnamon, to garnish

- Shake all the ingredients (except the garnish) in a cocktail shaker with ice. Strain into a cocktail glass. Garnish with a sprinkle of cinnamon.

JUNE

1 JUNE

THE RITZ PARIS OPENS (1898)

MIMOSA –

As a fifteen-year-old, César Ritz was told by his hotel boss that he would never make anything of himself in the hospitality business.

But Ritz, the last of thirteen children from a Swiss peasant family, defied these modest expectations, going on to become arguably history's greatest ever hotelier – and he even had a moreish cracker named after him.

Famed for his attention to detail and ability to keep high-class clientele content, he formed a powerful partnership with Auguste Escoffier, Europe's top chef at the time. Together, they opened London's Savoy Hotel and 'created' modern luxury where, over eight years, they (allegedly) misappropriated the equivalent of £2.5 million in pilfered wines and spirits, kickbacks and client 'entertainment' expenses.

Using this money, they opened the Ritz in Paris, where regular guests included Coco Chanel, Ernest Hemingway and (less popular) the Luftwaffe, who used it as their headquarters during the Second World War.

Here, in the 1920s, bartender Frank Meier reputedly created the Mimosa. One of the ingredients, Grand Marnier, was named by Ritz. Initially known merely as Marnier, Ritz tasted it and loved it so much he christened it 'Grand Marnier'. Meaning 'Big Marnier'.

~

12.5ml (½ fl oz) Grand Marnier | 40ml (1¼ fl oz) orange juice
120ml (4 fl oz) champagne

- Pour the Grand Marnier and OJ into a champagne flute and top up with champagne.

2 JUNE

BATTLE OF SANTIAGO (1962)

PISCO PUNCH -

Look at Chile on a map and you'll notice that its shape resembles that of a chilli. See?

No surprise then, that when Chile hosted the 1962 World Cup, things got a little bit spicy. The most violent tournament in World Cup history, the first two days saw four red cards, three broken legs, a fractured ankle and some cracked ribs.

And that was before the 'Battle of Santiago', a 90-minute fracas between Chile and Italy which seriously kicked off before kick-off. Prior to the encounter, just two years after a horrific Chilean earthquake left two million homeless, Italian journalists mocked the rubble-strewn nation and criticised the tournament's organisation. Then, just for good measure, they questioned the aesthetic qualities of Chile's womenfolk.

On gameday, at a Santiago stadium stuffed with 66,000 seething locals, the Chileans spat in Italian faces before English referee Keith Aston had even tossed the coin.

Within five tempestuous minutes, the game descended into a giant cartoon brawl – a massive *Beano*-esque cloud of dust with assorted fists, feet and colloquial expletives emerging from it at various angles.

Two minutes later, an Italian player was sent off for a dirty kick. When he refused to depart, the Chilean police got involved and pulled him off (oh, behave). Following more fouls and fisticuffs, another Italian was dismissed (but tried to sneak back on).

The second half saw more scuffles, spitting, scything tackles and general shit-housery, with Chile triumphing 2–0 against nine Italians – but not before a broken nose, three more police interventions and, after the weary Aston blew the final whistle, a peach of a left-hook to the jaw.

~

Serves 11 (or 9, if you're Italian)
70cl Pisco | 250ml (8½ fl oz) pineapple juice
500ml (16 fl oz) sparkling water | 250ml (8½ fl oz) lemon juice
Large piece of ice | Ice cubes | Pineapple slices in syrup, to garnish

- Pour all the liquid ingredients into a punch bowl over a large piece of ice. Serve in cups over ice and garnish with slices of sugar syrup-steeped pineapple.

---

3 JUNE

World Bicycle Day

45 DAYS OF HELL(ES), TO ØL –

With more than two-thirds of its citizens cycling to work or school, Copenhagen is regularly voted the world's greatest cycling city.

It has a great beer scene, too. One of the city's creative brewers is To Øl, meaning 'two beers'. A 'ToOl' is also something you use to fix your bike. Their thirst-slaking Helles is great gear after a big ride.

---

4 JUNE

Shane Warne's Ball of the Century (1993)

JIM BARRY 'COVER DRIVE' CABERNET SAUVIGNON, 2017 –

Over the course of his career, cricketer Shane Warne did outrageous things with his balls, both on and off the pitch.

None greater, however, than the 'Ball of the Century' he bowled during a Test match between England and Australia in 1993. On the second day at a gloomy Old Trafford, the slightly tubby, relatively unknown bleach-blond Australian took the ball for the first time in an Ashes Test.

Facing an even tubbier Mike Gatting, a batsman with over 40 Ashes innings under his overstretched belt, and an adept player of spin, Warne strolled up, wrist cocked, and zipped a rapidly spinning leg-break across Gatting's eyeline.

It landed way outside leg stump on a worn patch of pitch, then span sharply back across Gatting's pads, bat and considerable girth. As Gatting played his classic downward defensive shot, the turn of the ball, rotating at a ridiculous rate, bamboozled him, gliding past his bat and clipping his off stump.

A stunned Gatting spent several seconds gawping at the pitch. 'It was as though someone had just nicked his lunch,' recalled Graham Gooch, England's captain. 'If it had been a cheese roll, it'd have never got past him.'

The 'Ball of the Century' instantly revived the art of spin bowling after years of pace bowling, and marked the beginning of Australia's Ashes dominance – with Warne losing just one series in his entire 23-year Test career.

---

5 JUNE

National Danish Day

DANISH AKVAVIT –

Have you got a hygge knob? We don't like to boast, but we do – and today, it's turned all the way up to eleven.

It may be June, but we're up to our necks in Nordic knitwear amid the flickering light of scented candles, we've a Danish pastry in one hand and a cup of rosehip tea in the other, and we've replaced all the sports autobiographies on our bookshelf with a retro hipster typewriter.

To achieve the highest form of hygge requires akvavit, Denmark's national drink. It's a spirit distilled with botanicals, most notably caraway and dill. Cardamom, anise, fennel, coriander, lemon or orange peel are also occasionally added. Some liberal Danes even like it with cumin.

6 JUNE

SEX AND THE CITY AIRS FOR THE FIRST TIME (1998)

COSMOPOLITAN –

Sex and the City, which aired between 1998 and 2004, was a hugely successful comedy drama about four wealthy, independent, successful white women whose New York lives revolved around shoes, sex, shopping, dinners and, let's be honest, a preoccupation with men. Not necessarily in that order.

As a feminist portrayal of womanhood it certainly had its flaws and, looking back, the lack of racial, social and sexual diversity doesn't stand up today. But it did catalyse the global success of the Cosmopolitan cocktail: easy on the eye, easy to make and even easier to drink – plus the flamed orange peel adds drama to its creation.

Talking of flames, the Cosmo contains cranberry juice – which is great for cystitis.

~

40ml (1¼ fl oz) lemon vodka | 15ml (½ fl oz) Cointreau
10ml (½ fl oz) lime juice | 20ml (¾ fl oz) cranberry juice
Ice | Orange peel, to garnish

- Shake all the ingredients (except the garnish) in a cocktail shaker with ice. Shake, then strain into a chilled cocktail glass. Flame the orange peel and add to the drink.

7 JUNE

BIRTH OF DEAN MARTIN (1917)

SCOTCH WHISKY WITH SODA –

In his effortless pursuit of cool, Dean Martin didn't dick around. Growing up in small-town Ohio, Martin left school at 15 and became mates with the Mob, delivering bootleg liquor to illicit speakeasies.

A funny, velvet-voiced, good-looking rascal, Martin began his showbiz career as a superb slapstick comic before switching to more serious roles, appearing in several films alongside fellow 'Rat Pack' rogues Frank Sinatra and Sammy Davis Junior.

'Dino' was a dapper drinker whose childhood contempt for Prohibition removed all respect for teetotallers.

Seldom seen without a Scotch in his hand, he deliberately dialled up his amiable drunk act, purposely slurring his speech and playing for laughs as the tipsy, glassy-eyed crooner. But he knew how to handle his drink, and favoured a light Scotch whisky with soda.

8 JUNE

GEORGE ORWELL'S *1984* PUBLISHED (1949)

BLACKSHORE STOUT, ADNAMS

George Orwell: one of England's most famous novelists. As well as *1984*, his ominous premonition of a dystopian future, he penned 'The Moon Under Water', an iconic essay about the perfect yet mythical English pub, and *Animal Farm*, an allegorical tale featuring Communist pigs and Muriel, a clever goat that can read.

Orwell spent lots of time in Southwold on the Suffolk coast, home of the iconic Adnams brewery since 1872. They make great beer (and some superb spirits), including Blackshore Stout, a delicious, velvety 'Black Beauty' of a beer that's darker than an Orwellian nightmare.

9 JUNE

BIRTH OF PETER THE GREAT (1672)

RUSSIAN VODKA –

If you stacked a set of wooden Russian dolls featuring past tsars in order of political significance, Peter the Great would be the biggest – with all the less-important tsars inside his wooden tummy.

Standing at six foot seven, he was colossal in every way – even before putting on one of those big furry Russian hats. During his 53-year reign, he successfully expanded the Russian Empire, created St Petersburg and, inspired by the Enlightenment, founded the cultural and administrative institutions on which modern Russia is built.

It wasn't all work, though. He organised dwarf weddings and ordered them to jump out of pies, kept a pet monkey on the back of his throne and trained bears to fetch him his daily vodkas.

He'd also dish out vodka to peasants from a bear-drawn sleigh, and designed a triple-distillation device that was instrumental in improving the purity of Russian vodka.

He was also called the 'Antichrist' – but, hey, no one's perfect.

10 JUNE

FIRST UNIVERSITY BOAT RACE (1829)

FULLER'S VINTAGE ALE –

Every spring, hordes of genteel folk gather along the River Thames to witness the Cambridge vs Oxford boat race, an elitist hangover from an idealised, sepia-tinted English past.

Back in the day, the university crews consisted of genuine students: 'bloody good blokes' called Hugo with high hair who were studying Theology and that kind of thing.

But these bumbling crab-catching poshos have since been

usurped by elite international 'professional' rowers with killer abs and bulging biceps studying spurious PhDs in 'leisure management'.

No one really cares who wins – it's just an excuse for a few pints by the river. One of the best places to see these strapping young men pumping back and forth with their tiny little cox is from the historic Fuller's Brewery in Chiswick, home to this port-like Vintage Ale.

Buy two bottles. Drink one and cellar the other. When it increases in value, you can sell it, join the English elite and send your children to Oxbridge. Even the thick ones.

---

11 JUNE

Escape From Alcatraz Prison (1962)

BREW FREE! OR DIE IPA,
21ST AMENDMENT BREWERY –

Between the 1930s and 1960s, Alcatraz Prison was America's toughest federal penitentiary institution.

Situated on an island two kilometres from San Francisco's coast, it was almost impossible to escape from. But that didn't stop three armed robbers attempting the feat: brothers John and Clarence Anglin, and Frank Morris; who had a history of absconding high-security institutions alongside an absurdly high IQ.

After spending 3 months excavating escape tunnels through their cells' air vents using sharpened spoons, fashioning a crude raft from prison-issue raincoats and creating papier-mâché heads with soap and human hair from the prison's barbershop, they made a break for it.

After shinning up drainpipes, scrambling across the prison rooftop, cutting through the perimeter fence and shuffling down the embankment to the water's edge, they were never seen again.

As fragments of their raincoat raft washed up on the city's shore, they were presumed drowned. But no bodies have been found, and they're still on the FBI's 'Most Wanted' list. In 2016, a photo from 1975 emerged featuring two men in Brazil who resembled the Anglin brothers.

12 JUNE

RAIDERS OF THE LOST ARK PREMIERES (1981)

JOHNNIE WALKER BLACK LABEL –

There's a lot to like about Steven Spielberg's epic action-adventure classic. There's the boulder scene with the hat, Indiana Jones (Harrison Ford) shooting a big sword-wielding dude, a Nazi-empathising capuchin monkey in a waistcoat, and a cracking clothes-hanger gag.

There's even a great bar fight scene, when Ford knocks out an aggressive goon with a Johnnie Walker bottle. Ford received a Golden Globe Award for his performance, but critics shamefully overlooked the Nazi-empathising capuchin monkey in a waistcoat.

13 JUNE

BIRTH OF MALCOLM MCDOWELL (1943)

HORSFORTH PALE, HORSFORTH BREWERY –

In Stanley Kubrick's film *A Clockwork Orange*, Alex (played by Malcolm McDowell), eyeballs the audience while slowly sipping milk in the Korova Milkbar.

This isn't standard semi-skimmed stuff, though. This is Moloko Plus, a milk 'cocktail' laced with a shiver-inducing blend of barbiturates that, says Alex, 'makes you ready for a bit of the old ultra-violence'.

All rather unseemly. So, instead, have a nice pale ale from Horsforth, where McDowell was born.

14 JUNE

Michael Jordan wins NBA Championship with last shot for Chicago Bulls (1998)

RON ZACAPA CENTENARIO SISTEMA SOLERA 23 RUM –

Michael Jordan's last shot for the Chicago Bulls is hailed as one of the NBA's finest ever.

With Chicago trailing and just 5.2 seconds left, Jordan sunk a 20-foot jump shot to give them an 87–86 lead: their third straight championship, and sixth in eight years. It reaffirmed Jordan's status as the great basketball player of all time. After the game, the Bulls retired his number 23 jersey.

Jordan went on to launch his own tequila, but it's very expensive. So, instead, we're choosing this stunning Guatemalan rum. It shares Jordan's number (23) and is also slow-aged in oak barrels at high altitude – where 'His Airness' spent a lot of his time.

15 JUNE

Birth of Monsieur Mangetout (1950)

CHARTREUSE –

Michael Lolito was just nine years old when he gleefully wolfed down the shattered shards of a broken glass.

It was the first signs of pica, a disorder that makes you eat things that really shouldn't be eaten. Wholeheartedly embracing the ailment, Lolito become a novelty act known as Monsieur Mangetout ('Mr Eat Everything') in 1966.

Over a thirty-year 'career', he reckons he munched nine tonnes of metal, including: 18 bicycles, 7 TV sets, 6 chandeliers, 15 supermarket trolleys, a pair of skis, 2 beds, a computer and a Cessna light aircraft (which took two years to consume). He also ate a coffin.

He washed down his 'food' with mineral oil, but we suggest Chartreuse, a glorious green monastic liqueur made in the Alps near Mangetout's home town of Grenoble.

Containing 130 herbs and spices, and aged for five years in oak casks, it's used by local farmers to cure their cows' flatulence – which can't be as bad as Monsieur Mangetout's.

---

16 JUNE

BIRTH OF HARRY MACELHONE (1890)

WHITE LADY –

Harry MacElhone, among history's great bon-vivant bartenders, was a dry, dapper drink-smith from Dundee.

He first cut his cocktail teeth at the New York Bar in Paris on 5 Rue Daunou, which he bought 12 years later and renamed, rather unoriginally, 'Harry's New York Bar'.

MacElhone rose to fame during the roaring twenties. Hordes of Americans fled Prohibition for Europe, where everybody wanted to forget the horrors of war – and forget quickly.

MacElhone had just the thing: the White Lady, a punchy, potent cocktail capable of hazing one's memory from a hundred yards. It remains the drink for which he is renowned.

~

50ml (1½ fl oz) gin | 25ml (¾ fl oz) Cointreau
25ml (¾ fl oz) lemon juice | 1 egg white
Ice | Lemon zest twist, to garnish

- Shake the ingredients (except the garnish) in a cocktail shaker with ice, then strain into a glass. Garnish with a lemon zest twist.

17 JUNE

Roberto Cofresi Born, 1791

PINA COLADA –

Considered the last true Pirate of the Caribbean, Roberto Cofresi is known in his native Puerto Rico as the 'Robin Hood' of pirates – not because he had a fat mate, but rather because he stole pieces of eight from European ships and lavished his loot upon the locals, especially women, children and the elderly.

Notoriously ruthless in battle, he allegedly killed 400 men with his own hands and, an infamous philanderer, entertained countless women while they sat on his boat.

He also possibly created the Pina Colada, serving his weary crew a concoction of pineapple, coconut and white rum. Is this true? It is if you want it to be.

~

50ml (1½ fl oz) white rum | 75ml (2½ fl oz) pineapple juice
25ml (¾ fl oz) cream of coconut | 25ml (¾ fl oz) single cream
300g (10oz) cracked ice | Pineapple slice and cherry, to garnish

- Blend all the ingredients (except the garnish) in a blender with cracked ice, then pour into a hurricane glass. Garnish with a pineapple slice and cherry.

18 JUNE

National Icelandic Day

REYKA VODKA –

For a nation with a population of a little more than 350,000, Iceland's list of impressive achievements is not to be Bjork-ed at.

The Icelandic elected the world's first female president and the first openly gay head of government, swiftly jailed bankers following the global financial crisis in 2008, and, at the European Football Championships in 2016, they famously gave us all the clap.

This premium volcanic vodka, distilled using pure water filtered through 4,000-year-old molten rocks, is lava-ly.

19 JUNE

International Martini Day

GIN MARTINI –

No other drink captures the cocktail hour more magnificently than the Martini.

Rightly considered the king of classic cocktails, it's staggeringly simple, yet deceptively difficult to master. Unashamedly brazen in its booziness, transparent in appearance and intention, clean-lined yet complex, it's a V-shaped salute to the trudge and tyranny of modern life.

We recommend you keep it cold. As cold as possible. The shaker, the gin, the vermouth and the glasses. Everything.

~

60ml (2 fl oz) gin | 3 tsp dry vermouth
Ice | Olive, to garnish

- Stir the gin and vermouth in a mixing glass with ice, then strain into a chilled Martini glass. Garnish with an olive.

20 JUNE

The Earliest Day for Summer Solstice

HERNO GIN AND TONIC –

Let's go to the top of Sweden, where the sun stands still. To celebrate, locals go midnight skinny-dipping after a busy day of boogying around maypoles, stuffing their unfeasibly attractive faces with pickled herring and constructing flatpack furniture.

The Swedes traditionally sip schnapps, but here's an awesome, award-winning, juniper-led gin from the world's northernmost distillery on Sweden's 'High Coast'.

Garnish with a sun-like orange slice.

---

21 JUNE

NATIONAL GIRAFFE DAY

A PAIR OF HIGHBALLS –

Highballs? Giraffe? Dyageddit? Oh, never mind.

The highball is a simple, two-ingredient mix of spirit and something bubbly. Here's a brandy and soda highball for you.

~

Ice | 45ml (1½ fl oz) brandy | Soda, to top up
Orange slice, to garnish

- Fill a highball glass with ice cubes. Add the brandy, stir for 5 seconds, then top up with soda. Stir softly for 3 seconds. Garnish with an orange slice.

---

22 JUNE

MARADONA'S 'HAND OF GOD' (1986)

FERNET-BRANCA AND COCA-COLA –

Just four years after the Falklands War, with relations far from harmonious, England and Argentina met in the 1986 World Cup quarter-finals in Mexico.

Diego Maradona restored pride to his nation with the most infamous goal in World Cup history, single-handedly ending England's hopes in front of 114,000 fans at the Azteca stadium.

England's agricultural defence spent the first 45 minutes kicking the living daylights out of the Argentinian captain, but then, six minutes into the second half, Maradona surged towards the English penalty box.

As he flicked the ball to his right, his teammate Jorge Valdano miscontrolled it. England's Steve Hodge then sent a clumsy slice looping high into the area behind him.

As the ball dropped down towards the penalty spot, the race was on between the five-foot-four Argentinian and England's six-foot goalkeeper, Peter Shilton. Maradona leapt like a salmon, Shilton leapt like a badger, and – fist clenched close to the side of his head – Maradona poked the ball into the empty net.

As enraged England players ran towards the Tunisian referee, slapping the top of their hands as if asking for the time, Maradona celebrated wildly – with nothing more than a cheeky glance over his shoulder to see if he'd been rumbled.

He hadn't. None of the officials had seen the 'Hand of God', blinded either by divine intervention or, more likely, the sheer cajones required to pull off such an act.

Minutes later, things got seriously celestial when Maradona scored the greatest goal in World Cup history, weaving his way through a wilting England defence from his own half.

His two goals epitomised the two halves of Maradona's character: part genius, part outrageous swindler, loved and loathed in equal measure. Watch them both while enjoying Argentina's most popular drink, Fernet-Branca: a rich Italian bitters made with 27 different herbs and spices. Argentinians enjoy it with Coke – which, apparently, is exactly how Maradona liked a lot of things.

~

Ice | 35ml (1 fl oz) Fernet-Branca | Coca-Cola, to top up

- Fill a chilled highball glass with large ice cubes. Add Fernet-Branca and top up with Coca-Cola, poured at a 45-degree angle to avoid excess fizz.

23 JUNE

BIRTH OF ALAN TURING (1912)

ENIGMA IPA, WILD CARD BREWERY –

If it wasn't for the magnificent Alan Turing, we'd be writing this in German – which would be a pain, as we can't speak it, let alone write it.

A code-breaking mathematic genius and unconventional war hero, Turing created a device that deciphered the German Enigma machine and was instrumental in the Allied victory.

But Turing was shamefully turned upon by his country and prosecuted for homosexuality in 1952. As an alternative to prison, he accepted chemical castration, before killing himself two years later. A half-eaten apple dipped in cyanide was discovered next to his deathbed.

Only officially pardoned in 2013, the quite brilliant Turing was an eccentric who, according to Bletchley Park co-workers, cycled around the grounds while wearing pyjamas and a gas mask – the latter to help with his hay fever. When required in London, he would often run the 40 miles in his suit, as it was quicker than waiting for a lift. And, in an office crammed with the finest code-breakers in the country, he chained his favourite coffee mug to the radiator by his desk using a combination lock.

No one tried to steal it, though, as no one wanted to be chased for 40 miles by a mathematical mastermind wearing pyjamas and a gas mask.

Have this cracking IPA in a mug – it's what he would have wanted.

24 JUNE

Ice Cold in Alex PREMIERES (1958)

CARLSBERG –

At the end of *Ice Cold in Alex*, Captain Anson is desperate for a f***ing pint.

He's a broken man: an alcoholic exhausted by the barbarism of war and brutalised by the baking North African heat, he's just endured a highly dangerous Dante-esque journey across the desert to Alexandria in a knackered ambulance called 'Katy' – along with two nurses and a sergeant.

It may sound like a sketch from Benny Hill, but it wasn't. There were minefields, breakdowns, and one of the nurses was shot by a German patrol. He also had to give a shifty South African a lift. All in all, shit times.

But now Captain Anson, his mission successfully completed (notwithstanding the dead nurse, of course), is sitting on a bar stool in Alexandria, his eyes stinging with sweat, staring at a gleaming, frosted glass of ice-cold lager.

Having been hoodwinked by mirages in the desert, he tenderly traces his finger down the side of the arched glass, leaving a trail in the dew. He waits, savouring the moment, before grabbing the glass, plunging his face into the froth and gulping it down in one, his leathery gullet pulsating in pleasure.

He then places the empty glass on the bar and mutters, 'Worth waiting for.'

25 JUNE

THE RAINBOW FLAG FIRST FLOWN AT SAN FRANCISCO GAY FREEDOM DAY PARADE (1978)

RAINBOW ROAD –

The Rainbow Flag, the global symbol for gay rights, was the brightly coloured brainchild of Gilbert Baker, a former military man turned drag queen who moved to San Francisco at the height of the gay rights movement in the 1970s.

Commissioned by Harvey Milk, California's first openly gay elected official, the flag was first flown in 1978 at the Gay Freedom Day Parade. It achieved worldwide prominence following the assassination of Milk later that year and, in 2015, two years before Baker's death, it was emblazoned on the walls of the White House to commemorate the legalisation of same-sex marriages.

~

25ml (¾ fl oz) vodka | 12.5ml (½ fl oz) watermelon liqueur
12.5ml (½ fl oz) apricot liqueur | 25ml (¾ fl oz) lime juice
10ml (½ fl oz) passion fruit syrup | Ice

- Shake all the ingredients in a cocktail shaker with ice cubes, then strain into a tall glass filled with crushed ice. Garnish with a rainbow flag.

26 JUNE

JFK DECLARES HIMSELF A BERLINER

BERLINER WEISS –

John F. Kennedy's '*Ich bin ein Berliner*' speech is a masterful piece of political rhetoric to rival Abraham Lincoln's transcendentalist 'Gettysburg Address', Lyndon Johnson's 'We Shall Overcome' civil rights sermon and, of course, Donald Trump's disinfectant diatribe at the height of the COVID-19 crisis.

Delivered in 1963 in West Berlin, in front of a crowd of more than 120,000, JFK's anti-Communist address confirmed America's solidarity with West Germany, less than two years after Soviet-occupied East Germany constructed the Berlin Wall.

In an impassioned address he'd re-written after finding the original too sympathetic to the Soviets, Kennedy said: 'Today, in the world of freedom, the proudest boast is "*Ich bin ein Berliner*!" ... All free men, wherever they may live, are citizens of Berlin, and therefore, as a free man, I take pride in the words "Ich bin ein Berliner!"'

Berlin has its own beer style: Berliner Weiss, a face-contorting traditional tart wheat beer laced with funky bacteria called *lactobacillus* that convert sugars into lactic acid.

It's often served with fruit syrup to soften the sourness, and was popular in the 19th century (when it was dubbed 'the Champagne of the North' by Napoleon).

27 JUNE

WORLD'S FIRST EVER CASHPOINT

MILLION DOLLAR COCKTAIL –

In 1965, British inventor John Shepherd-Barron was in the bath, watching his floaty bits poking through the bubbles, when he came up with the idea of an 'automated teller machine'.

He told the big boss of Barclays Bank about it over a pink gin and, within two years, the world's first cash machine had been installed on a high street in Enfield. It was, with hindsight, a bit rubbish. As there were no debit cards in 1967, it was based on a convoluted voucher system that required customers to queue up at the same window where they'd otherwise withdraw cash. It repeatedly ran out of notes every weekend and the vouchers were also slightly radioactive.

Still, in honour of the bathtub architect of the ATM, we're making ourselves a Million Dollar cocktail with pink gin.

~

50ml (1½ fl oz) pink gin | 25ml (¾ fl oz) vermouth
15ml (½ fl oz) pineapple juice | 7.5ml (¼ fl oz) grenadine
½ egg white | Ice | Orange zest twist, to garnish

- Shake all ingredients (except the garnish) in a cocktail shaker with ice. Strain into a cocktail glass and garnish with an orange twist.

---

28 JUNE

Birth of Peter Paul Rubens (1577)

DE KONINCK FROM A BOLLEKE –

Peter Paul Rubens was a big-deal Baroque painter famed for his buxom nudes and expansive strokes.

Just like rapper Sir Mix-a-Lot, whose appreciation of more rotund female behinds is renowned, Rubens wasn't lying when he proclaimed his passion for painting women with plump posteriors. 'I paint a woman's big, rounded buttocks,' he said, 'so that I want to reach out and stroke the dimpled flesh.'

Big butts and bosoms were by no means the only thing Rubens painted during his remarkably prolific career – he also did monarchs wearing detailed ruffs, portraits, landscapes, depictions of mythological and allegorical subjects, hunting scenes and altar pieces. But not skirting boards.

His claim that 'I'm just a simple man, standing alone with my old brushes, asking God for inspiration' was a humble brag, as his political and artistic influence stretched far beyond his canvas. Rubens was a deft diplomat, skilled statesman and a classically educated humanist scholar who was knighted in both England and Spain.

Rubens' most famous studio was in Antwerp, which is home to the wonderful De Koninck Brewery, whose flagship beer is drunk from a glass called a Bolleke. No, really. Sadly, as De Koninck was founded 200 years after Rubens' death, no one knows whether he was as good at painting Bollekes as he was at painting women's arses.

---

29 JUNE

Bentley Blower Sold for £5 Million (2012)

THE BENTLEY COCKTAIL –

Daring, dashing and debonair, the Bentley Boys were a group of fast-living, fast-driving, champagne-swilling, lady-killing, cravat-wearing, sooty-faced aristocrats.

The Bentley Boys' *beau idéal* was the moustachioed Sir Henry 'Tim' Birkin, who frittered away the family fortune in his pursuit of pleasure around the tracks at Le Mans, Brooklands and Nuremberg.

Not interested in winning races, Birkin simply craved speed and approached Bentley's owners to create a super-fast 4.5-litre car called the 'Blower' to compete with European rivals.

When Bentley's boss declined, Birkin plunged his last pennies into the project and, ever the charmer, persuaded his posh pals to do the same. Ridiculously rapid, and with Birkin at the wheel, the Bentley Blower reached a blistering record speed of 137.96mph.

In the debut James Bond novel, *Casino Royale*, the Blower was 007's first car. Birkin famously drove it up the stairs at the Savoy Hotel in London, where bartender Harry Craddock first mixed the

Bentley cocktail in the Bentley boys' honour.

In 2012, 83 years after it was first unveiled, Bonhams auctioneers sold it for £5,042,000.

~

50ml (1½ fl oz) calvados | 50ml (1½ fl oz) Dubonnet Rouge
3 dashes of orange bitters | Ice

- Combine the ingredients in a mixing glass over ice and stir until chilled. Strain into a chilled Nick & Nora glass.

---

30 JUNE

CHARLES BLONDIN CROSSES NIAGARA FALLS ON A TIGHTROPE (1858)

HIGH WIRE WEST COAST PALE ALE, MAGIC ROCK BREWING –

Back in 1858, in front of 25,000 anxious onlookers, a 34-year-old French acrobat walked across Niagara Falls on a 396-metre (1,300-foot) rope 5cm (2in) in diameter, holding a 6-metre (20-foot) pole.

Known as the Great Blondin on account of his golden locks, the five-foot-five daredevil wore pink tights covered in sequins and light leather shoes, with nothing to save him from plummeting into the misty, swirling waters below. Expectation of disaster, claimed Blondin, would simply incite it.

As he approached the midway mark, Blondin sat down on the sagging rope, with the Maid of the Mist tourist boat below him. Ever the showman, he lowered down a rope and whipped up a bottle of wine from the deck.

After taking a couple of swigs, he stood up and kept going, his strutting step breaking into a sprint as he approached Canada. Over the following years, he repeated the feat wearing stilts, a blindfold and a gorilla costume (not at the same time). He even did it pushing a wheelbarrow, presumably to help carry his enormous 'Niagaras'.

1 JULY

SERGEANT PEPPER'S LONELY HEARTS CLUB BAND REACHES NUMBER ONE IN THE US CHARTS (1967)

SALTY DOG –

Dogs probably don't like The Beatles' *Sergeant Pepper's Lonely Hearts Club Band*. The band recorded a high-frequency whistle that runs after the final chord of the last song on the record, 'A Day in the Life'. Only dogs can hear it, and it probably winds them up a bit. Otherwise, it's a pretty good album.

~

Pinch of rock salt, to rim the glass | Ice
50ml (1½ fl oz) gin or vodka | 100ml (3½ fl oz) pink grapefruit juice
10ml (½ fl oz) sugar syrup

- Rim a highball glass with salt and fill with ice. Shake the remaining ingredients in a cocktail shaker with ice, then strain into the prepared glass.

2 JULY

AMELIA EARHART DISAPPEARS (1937)

BÉNÉDICTINE –

After being celebrated as the first woman to fly alone across the Atlantic Ocean, Amelia Earhart's attempt to make it around the world ended in disaster when her plane crashed into the Pacific.

By 1939 it was assumed she had drowned, and her body was never discovered, but more recently scientists have suggested a skeleton unearthed on the island of Nikamuru might fit her profile. Which isn't particularly flattering, but does at least prove the woman had a spine. Apparently, there was also a woman's shoe found nearby. We're no Poirots, but that seems like a useful clue.

More importantly, they discovered an empty bottle of Bénédictine in the vicinity. Bénédictine was marketed as an elixir that brought cheer to virtuous and enduring monks, so the citrus-sweet liqueur, complete with 56 herbs, roots, fruits and spices, seems a fitting sip for any last-gasp efforts at survival on a desert island.

---

3 JULY

BENZ UNVEILS HIS NEW CAR (1886)

TANQUERAY ALCOHOL-FREE 0.0%

When mechanical engineer Karl Benz designed his first automobile in Mannheim, Germany, he experimented with alcohol as fuel, but sadly for the planet, petroleum won out. As well as being a planet-killer, you can't drive to the pub and enjoy a drink, so cars are a bit shit really. Not that you're obliged to drink alcohol in the pub – and, if you are driving, there are now plenty of grown-up non-alcoholic drinks. Using centuries of distilling expertise, Tanqueray has put some of its gin genius into this non-alcoholic gin, Tanqueray 0.0., and a big whack of juniper ensures it stands up in a tonic.

4 JULY

Independence Day (US)

SIERRA NEVADA PALE ALE –

It seems fitting to celebrate an independent brewery who defeated the ruling overlord of fizzy, flavourless lager. That brewery was Sierra Nevada, one of the biggest successes of America's craft-beer revolution.

Creator Ken Grossman started out with a home-brewing store in Chico. After selling this business, he invested in a couple of dairy tanks and a soft-drink bottler to start his ramshackle brewery. The unspectacular tools belied Ken's fabulous flagship beer, a lip-smacking pale ale with a distinctive hop character that inspired an endless array of hop-tastic American beers.

A move to state-of-the-art facilities means Sierra Nevada is now available in every state, and indeed across the world. But familiarity has not bred contempt among craft-beer drinkers, who, along with many mainstream drinkers, have adopted it as their 'go-to' beer.

5 JULY

Battle of Wagram Begins (1809)

CHAMPAGNE –

Wagram saw Napoleon take on the Austrians in a hugely bloody battle that lasted two days, with tens of thousands of deaths on both sides. The most notable casualty was arguably 19th-century French cavalry man, Hussar Antoine-Charles-Louis Comte de Lasalle, an exceptional drinker who fell on the second day of fighting.

Lasalle was a master of the *sabrage* technique, the rather impressive uncorking of a champagne bottle with a sabre. Far from over-compensating for other phallic failures, the swordsman's swagger was very much part of a top-to-bottom dapping, derring-do.

Sporting dazzling cherry 'Look At My F*cking Red Trousers' and adorning his jacket with abundant accoutrements, the gung-ho Hussar General was a very flashy flash in the pan. He once insisted that any soldier who wasn't dead by 30 was dishonourable, and went out in his own blaze of glory, aged 34. Riding into the fight knowing defeat was inevitable, he galloped to the front with his bottle-opening sabre securely in its hilt, instead raising his pipe before being engulfed in the discharge of countless weapons.

---

6 JULY

FEDERER VS NADAL (2008)

STRAWBERRY MARTINI –

When Nadal beat Federer 6–4, 6–4, 6–7, 6–7, 9–7 during a game with lots of running around and not a single lollipop serve, many fans argued it was the greatest final of all time.

But they forget about the 1936 final, when Fred Perry beat Gottfried von Cramm, 6–1, 6–1, 6–0 in 40 minutes – and, although this one happened on 2 July, we want to draw attention to it.

For a start, this final was over really quickly, which freed up a Sunday to focus on barbecue meat and Martinis. But more importantly, it allows us to talk about Cramm.

Cramm was an ace guy on the court, but a more impressive figure off it. One of Germany's great players, he was gay and despised the Nazis, eventually being imprisoned for his sexuality and for supporting Jews. Upon his release, he was forced to fight on the Eastern Front, and, despite his strong objections to the regime, he defended his troops and was awarded an Iron Cross for bravery. He was even involved in the failed July Plot to assassinate Hitler. All of which kind of pisses on Rafa and Fed's chips a bit.

~

Vermouth, to rinse the glass | 6 strawberries, plus 1 to garnish
60ml (2 fl oz) vodka | 10ml (½ fl oz) sugar syrup | Ice

- Rinse a chilled Martini glass with vermouth. Muddle the strawberries in a shaker, then add the vodka and syrup. Shake hard with ice, then fine strain into the glass. Garnish with a strawberry.

---

7 JULY

Birth of Tamara Mellon (1967)

COBBLER –

'If the shoe fits,' Tamara Mellon might have said when setting up her footwear company with co-founder Jimmy Choo, 'then someone is going to make a lot of money.' And thus she did. Tamara produces fancy shoes, so try a fancy cobbler with Chivas-blended Scotch.

~

Ice | 60ml (2 fl oz) Chivas-blended whisky
15ml (½ fl oz) Grand Marnier | 15ml (½ fl oz) cognac
Orange slice and mint leaf, to garnish

- Fill a wine goblet or rocks glass with crushed ice. Shake the whisky, Grand Marnier and cognac in a cocktail shaker with ice, then strain into the glass and garnish with an orange slice and mint leaf.

---

8 JULY

Spice Girls Release Debut Single 'Wannabe' (1996)

SPICED MULE –

Scientists have proved that women possess a heightened human oral perception – or, in lay people's terms, a more sophisticated sense of taste – than men, making them excellent supertasters in the world of spirits.

Girl power.

Which brings us neatly on to the Spice Girls, whose debut single 'Wannabe' was so successful it broke records set by The Beatles.

~

50ml (1½ fl oz) spiced rum | Ice
½ lime, plus a wedge to garnish | 150ml (5 fl oz) ginger beer

- Pour the rum into a highball glass over ice, then squeeze in the lime juice and top with ginger beer. Garnish with the lime wedge.

---

9 JULY

ZIDANE HEADBUTT (2006)

RICARD PASTIS –

Zidane might have been a two-time World-Cup winner if he hadn't head-butted Italian defender Marco Materazzi towards the end of his second final. Sadly, there's no room for ifs when there are butts, and once he was sent off, Italy won the game.

Zidane hails from Marseille, as does Pastis, an anise-scented aperitif originally designed to fill the void when absinthe was banned for making people delirious. *NB: Absinthe doesn't make people delirious*. Pastis replicates the flavours of its stronger forebear and the ritual of its serve, using one part spirit to five parts ice-cold water.

---

10 JULY

BORIS YELTSIN SWORN IN AS RUSSIAN PRESIDENT (1991)

VODKA –

Yeltsin's role in Russian history was to direct his nation towards democracy and establish a new market economy, but his heavy drinking and ham-fisted flailing steered it careening like a malfunctioning Lada into mass unemployment and a hyperinflation hangover.

Within just two years of his election, Yeltsin had let rip such a potent wind of political despair his rivals launched a failed referendum on his leadership.

This warning shot wouldn't abate his drinking, though. For example, when he met Kyrgyzstan president Askar Akayev, Yeltsin drank vodka so heavily he played the spoons, percussion-style, on Akayev's bald head.

When in Stockholm, he drank excessively and compared Bjorn Borg's face to meatballs, then declared massive cuts in the Russian nuclear stockpile and almost fell off the stage.

And in 1995, while a guest at the White House, he drank too much, snuck past Secret Service body guards and hailed a taxi in the street in a bid for pizza, all while wearing only his underpants.

In Yeltsin's defence, converting Russia to a democracy must've been pretty stressful, but it might have been easier to do it sober.

---

11 JULY

CHARLOTTE COOPER WINS OLYMPIC GOLD (1900)

SATURN COCKTAIL –

Tennis pioneer Charlotte Cooper was one of the first women to serve over arm and was respected for her exceptional serve-and-volley game. At age 26, she lost her hearing but continued to play while totally deaf. Incredible stuff, and Cooper's record of eight consecutive Wimbledon finals lasted 90 years, until Martina Navratilova earned her ninth finals appearance in 1990.

'But what's this got to do with Saturn?' we hear none of you ask. Well, you have those Olympic rings, and Saturn has rings ... That's it. You're welcome.

~

45ml (1½ fl oz) gin | 15ml (½ fl oz) fresh lemon juice
7.5ml (¼ fl oz) passion fruit purée | 7.5ml (¼ fl oz) orgeat
7.5ml (¼ fl oz) falernum | Crushed ice
Lime slice and a cherry, to garnish

- Blend all the liquid ingredients in a blender with a cup of crushed ice, then pour into a rocks glass filled with crushed ice. Stir and top with a little more ice. Garnish with a lime slice and cherry.

---

12 JULY

POINT BREAK RELEASED (1991)

BIG WAVE GOLDEN ALE, KONA BREWING CO. –

Hang ten, ride rad barrels and don't be a dick dragger, dude. Surf was indeed up when Oscar-winning director Kathryn Bigelow brought to our screen a cult movie about an undercover FBI agent infiltrating a gang of big-wave-riders-cum-bank robbers.

All very gnarly, but we'll enjoy our wave on a sofa in our pants, if it's all the same, and it'll arrive in the form of Kona Brewing's Big Wave Golden Ale. Much like the movie, this light-golden blonde beer isn't too complicated, but there's a subtle hop profile, making it a fine beer for a beach barbecue – or, indeed, post-bank robbery.

Vaya con dios.

---

13 JULY

LIVE AID CONCERT (1985)

AMARULA –

Nearly 40 per cent of the world's population are estimated to have watched the Live Aid broadcasts, with gigs being performed at Wembley in London and John F. Kennedy Stadium in Philadelphia, among other nations. Granted, not many viewers would've been sipping Amarula at the time, but this African spirit uses real marula fruit, harvested directly from marula trees. The fruit is fermented and distilled before being aged for three years and blended with cream.

---

14 JULY

BASTILLE DAY

CHEAP FRENCH RED –

In 1789, the French properly lost their *merde* and stormed the Bastille. But, *tres interessant*-ingly, historians including 19th-century goody-*deux-chaussures* Hippolyte Taine claimed the bloodthirsty behaviour was actually due to intoxication, arguing the angry mob were simply drunk and disorderly.

Alcohol is often held responsible for misdemeanours of the masses, even when the masses are entitled to be *fromaged* off, as was the case in the 1780s. But Hippolyte neglects to mention that everyone, rich and poor, was drinking wine in France in the 18th century, and the taxes levied on alcohol paid for schools, hospitals and loads of other worthy bits and bobs.

Meanwhile, the revolutionaries' frustration over rising wine prices was also only part of the problem, there was as much consternation about the excessive tax on salt and bread. So, while the brawlers had a decent dose of Dutch courage in the system, the wine would've been less potent if they weren't so bloody famished.

Anyway, whatever the trigger, they did that Bastille building *right* over. Although this wasn't *that* impressive, since the former royal fortress and monument to the tyranny of the Bourbon monarchs was a shoddy wreck. But a fight is a fight, and the French are welcome to celebrate it.

A year later, the French toasted freedom with a wine-fuelled party in a massive tent pitched on the building site of the fallen Bastille. This time, they lined their stomachs and there was no boozy bust-up – although there were plenty of busts, thanks to a decision to symbolise liberty by running naked through Paris. We tried this last time we celebrated in Pairs on Bastille Day and can report, rather than freedom, this is now a symbol of being legally detained, and the prisons today are much more effective.

During the 1780s, the sans-culottes sank cheap *vin rouge*, and after the revolution they banned luxury *Grand Cru* wines from 1792–1795 as they sought a more egalitarian status quo. This provides you with the perfect excuse to turn up at a party with plonk.

---

15 JULY

DIE HARD RELEASED (1988)

YAMAZAKI DISTILLER'S RESERVE –

One of the most iconic deaths in movie history is the moment antagonist Hans Gruber falls to his doom in slow motion from the top of the Nakatomi Plaza skyscraper in *Die Hard*. The late but great Alan Rickman plays villain Gruber, while Bruce 'McLane' Willis is the hero, spending most of the movie in a vest. If you've not seen it, you really must. There's a real sense of social realism in the piece, and as with his previous work in *Predator*, director John McTiernan's authorial expressiveness delivers thought-provoking and welcome aesthetic pretensions.

While the Japanese Nakatomi Corporation is fictional, we recommend an excellent whisky from the Suntory Distillery. Distiller's Reserve is a single-malt with vibrant red berry notes to match Willis's blood-stained vest, it's matured in Japanese Mizunara oak, Bordeaux wine and sherry casks and bursting with the energy of the movie, but mercifully delivers a lot more subtlety.

16 JULY

CATCHER IN THE RYE PUBLISHED (1951)

SCOTCH AND SODA –

In this 1951 American classic, protagonist Holden Caufield orders a Scotch and soda, but when asked for ID is refused, so asks for a rum and Coke instead. It was an interesting approach, trying to twist the arm of a waiter by switching from Scotch to rum, but rum will always be booze, and Holden still looks underage so, unsurprisingly, it fails. He gets a Scotch and soda later in the book, though, and apparently author J.D. Salinger was a fan.

Salinger carried the manuscript for this book around with him during the Second World War, when he served in the 4th Counter Intelligence Corps detachment during the D-Day Landings, experiencing over 11 months of gruesome conflict. Returning safely, he enjoyed huge success with this novel, but due to rarely appearing in public, he was often described as a 'literary recluse'.

All we'd say about this 'recluse' thing is, perhaps he just didn't want to talk to strangers about the things he'd done or things he didn't want to do again. In our experience, other people can be really annoying, so maybe those who claimed he was a 'recluse' should have questioned whether they were the sort of people J. D. Salinger – or anyone else for that matter – needed to meet. He was probably just minding his own business and enjoying a Scotch and soda in peace. Fair play to him.

17 JULY

POTSDAM CONFERENCE BEGINS (1945)

VODKA OR WINE – /

During the 1945 post-war get-together between world leaders Churchill, Stalin and Truman, the Russians hosted a state dinner and put on quite the spread. Caviar, foie gras, suckling pigs and a variety of cheeses were all paired with bottomless glasses of wine and vodka.

The Russians insisted on 14 toasts with vodka while they dined, and Stalin led the charge, although later admitted he replaced vodka with wine because he'd endured a minor heart attack shortly before. Lightweight Truman struggled to keep pace, but Churchill poured the stiff drinks through his stiff upper lip and stayed the distance.

Worth adding, though, that as these smug and sozzled 'Big Three' leaders sweated and swelled like the suckling pigs on which they were gorging, they slowly slid the world into a protracted Cold War. It's also worth noting this was all happening while the Americans were secretly testing the first nuclear bomb. And that Churchill had already (albeit unknowingly) got the chop – to go with the chop he was stuffing in his face. And that Stalin was generally a rather nasty piece of work.

But it's important to stress that, once again, it was alcohol that oiled the wheels of peace negotiations.

---

18 JULY

JEAN VAN DE VELDE LOSES THE BRITISH OPEN (1999)

ARMORIK CLASSIC BRETON SINGLE MALT –

Some say golf can be a cruel sport, but they are wrong. While golf *is* 'flog' spelt backwards, there's nothing cruel about a bunch of moneyed characters sauntering around acres of uber-manicured grounds.

Even so, when the Frenchman Jean van de Velde bottled it on the last hole of the 1999 British Open, it was a tough watch. Leading by three strokes at the 18th, he could afford to hit a disastrous double-bogey six and still earn the Claret Jug and a place in golf history. As it was, he pinged shots into grandstands and water features, melting like Camembert as he recorded an unthinkable triple bogey, before losing in a play-off. *Mon dieu*, someone get that man a hanky for his tears – *and* all those bogeys.

Golf is not for the French who boast only one major winner, Arnaud Massy, winning the British Open in 1907. Nor is the nation recognised for its whisky, another famous export from the Scots. That said, in Brittany, you'll discover Warenghem, which has been turning out palatable single malts since 1998 and exports its Armorik globally. So, while it seems unusual, give their whisky a go: you'll have fewer regrets than Jean.

19 JULY

FIRST EPISODE OF *MAD MEN* RELEASED (2007)

HAMM'S BEER –

Don Draper drank. *A lot.* His standard selection was an Old Fashioned, but in general, he drank anything to hand, like some sort of gasping alcoholic fish.

So, much as we love *Mad Men*, don't drink like Don.

In one episode he benches the cocktails, opting for a Fielding beer instead. The brand was designed specifically for the show, but the can replicates the 1960s Hamm's Beer, which we assume is an in-joke, what with Don Draper being played by Jon Hamm.

Hamm's is an American cult classic. The brewery was established in 1865 in St Paul Minnesota. Now owned by Coors, the Premium, Golden Draft, and Special Light are all standard lager-style beers, without too much to upset the palate – which might be why Don skulls his like fizzy pop on his non-cocktail drinking day.

20 JULY

FIRST STEPS ON THE MOON (1969)

COMMUNION WINE –

History will rightly remember Neil Armstrong as the first man to walk on the Moon, but Buzz Aldrin was the real star. Buzz not only followed his fellow astronaut on to the bouncy surface, he also brought wine on the adventure. Granted it was for religious purposes – Buzz used the wine for Communion – but, as the first man to drink on the Moon, he earns a higher status than Armstrong in this book.

21 JULY

FINAL *HARRY POTTER* BOOK IN THE SERIES IS PUBLISHED (2007)

BUTTERBEER

When J. K. Rowling released the last book in the series – *Harry Potter and the Deathly Hallows* – it sold 11 million copies in a mere 24 hours. Now *that's* magic.

The eight adapted films went on to make $7.7 billion (at the last count). This makes Potter the third-highest-grossing film franchise, behind Marvel and *Star Wars*, but ahead of Bond (wizards beat spies, but not aliens and superheroes). There was also (at last count) $2 billion in DVD sales and $7.3 billion from merch.

Not a bad return for a story about a teenage wizard who cocked around a boarding school pulling wands from wizard's sleeves.

So, with all this moolah sloshing around, why not chip into the Potter pot by buying some Butterbeer. This is the butterscotch drink Harry and his school chums chug on in places like the wizard village Hogsmeade, and while the drink was once fictional, it is now magically licensed and available to buy. It's also alcohol-free, vegan, vegetarian and gluten-free, so suitable for a non-drinking day and can be shared with the kids as well. *Expelliarmus*, indeed.

---

22 JULY

AVENGERS ENDGAME BECOMES HIGHEST-GROSSING FILM OF ALL TIME (2019)

SUPERHERO COCKTAIL –

Welles, Bergman, Fellini, Truffaut – none of them saw it coming. Scorsese was so gutted he claimed it wasn't even cinema. But there it was, a film about superheroes earned $2.79 billion globally to become the biggest box-office hit in history.

Marco Corallo is a bartender, so he can't fly or shoot webs out of his hands, but he can make a decent drink. He originally created this cocktail using overripe bananas and discarded fruit juice to focus on food waste. Thus, in some ways, he is trying to save the world – just like Iron Man.

~

50ml (1½ fl oz) Bacardi 8-Year-Old Rum
15ml (½ fl oz) Amaro Ramazzotti
20ml (¾ fl oz) overripe banana purée | 25ml (¾ fl oz) mango purée
45ml (1½ fl oz) pineapple juice | 20ml (¾ fl oz) lime juice
2 teaspoons sugar | Ice | Pineapple leaves and lime shell, to garnish

- Shake all the ingredients (except the garnish) in a cocktail shaker with ice. Strain into an ice-filled tiki mug or Collins glass. Garnish with the pineapple leaves and lime shell.

23 JULY

One Direction Formed (2010)

ÆCORN BLOOD ORANGE SPRITZ

In an interview, singer Liam Payne revealed his pre-show ritual when performing with boy band One Direction was a vodka Red Bull, but he used a quadruple measurement of vodka mixed with the energy drink. He called it, rather cleverly, 'QuaddyVoddyRedBull'. The name isn't bad, but it'll give you the opposite of direction, and since some of the lads were not of drinking age when they formed, go for something much more discerning and non-alcoholic with this Æcorn Bitter cocktail.

~

Ice | 50ml (1½ fl oz) Æcorn Bitter
Fever-Tree Blood Orange Soda, to top up
Blood orange slice, to garnish

- Fill a wine glass with ice, pour in Æcorn bitter and then stir in the blood orange soda. Garnish with a blood orange slice.

24 JULY

Birth of Alexandre Dumas (1802)

ARMAGNAC –

'All for one and one for all, united we stand, divided we fall.' So wrote Alexandre Dumas in his smash hit *The Three Musketeers*. Dumas died in 1870, so wouldn't witness his story immortalised by cartoon *Dogtanian and the Three Muskahounds*, but he did get to enjoy Armagnac, the oldest French brandy.

Armagnac is the spirit of Gascony in southwest France, home of Dumas's main character D'Artagnan, and while we love cognac, it's worth saying the French have done a great job of keeping Armagnac all to themselves. Just 2.5 per cent of the 150 million bottles of cognac consumed worldwide are done so in France, while less than half of all the Armagnac sold every year, around 6 million bottles, is enjoyed abroad.

That's because it's smashing gear, truly lovely liquid. It can be ramshackle, rural and rustic, and yet as refined as the brandy from the manicured *maisons* that control cognac, making it '*La France Profonde*' in liquid form.

---

25 JULY

NATIONAL MERRY-GO-ROUND DAY (USA)

VIEUX CARRÉ –

The folks in the Byzantine Empire were a right laugh. Take their Hippodrome of Constantinople, a veritable fun house of public parades that boasted highly entertaining public executions and shaming of the emperor's enemies. One time – this was hilarious – they had an enemy publicly whipped while he was sent naked around the track riding backwards on a donkey. An ass, making an ass out of man with his ass out. And then there was the time they blinded people publicly, which sadly meant they couldn't see all the people giggling at them. Such a laugh.

Less violent were their merry-go-rounds. A 500 CE Asia Minor stone carving indicates that, in between blinding people, they got their jollies from harmless spinning baskets. That a merry-go-round can still entertain and delight us millennia later is certainly a reason to have an entire day dedicated to them.

The Hotel Monteleone in New Orleans has a Carousel Bar, which, much like the very best merry-go-rounds, does indeed go around. The 25 seats have circled the circus-clad bar once every 15 minutes for more than 40 years, as it serves up its signature 1930s creation, the Vieux Carré.

~

30ml (1 fl oz) rye whiskey | 30ml (1 fl oz) cognac
30ml (1 fl oz) sweet vermouth | 15ml (½ fl oz) Bénédictine
2 dashes of Peychaud's bitters | Ice | Lemon zest twist, to garnish

- Add all the liquid ingredients to a rocks glass and fill with ice. Stir briefly, then garnish with a lemon zest twist.

---

26 JULY

EINSTEIN'S GENERAL RELATIVITY THEORY TESTED (2018)

BLACK HOLE PORTER, OAKHAM ALES –

In 2018, the European Southern Observatory, the foremost intergovernmental astronomy organisation in Europe, peered through their appropriately named Very Large Telescope in Chile and observed the S2 star as it passed a black hole. Recording the position and velocity measurement, they reported results inconsistent with Newtonian predictions, but 'in excellent agreement with the predictions of general relativity', thus supporting Einstein's theorising on the movement of a star passing through the extreme gravitational field near a supermassive black hole in the centre of the Milky Way.

That's all we have to say about that carry-on, but it sounds important.

---

27 JULY

BIRTH OF WRESTLER TRIPLE H (1969)

DANCING GNOME TRIPLE LUSTRA –

This one is a gift: a wrestler called Triple H (real name Paul Levesque) to go with a triple IPA. Triple H won the WWF championship in 1999; meanwhile, Triple Lustra is brewed by

Pittsburgh craft specialists Dancing Gnome. The beer is hopped three times more than Dancing Gnome's revered pale ale, and bounces off the ropes with citrus flavours, smacking you down with an 11% ABV.

---

28 JULY

INDEPENDENCE DAY (PERU)

PISCO SOUR –

Peruvians claim they invented the Pisco Sour, but so do the Chileans. Sadly, Chile's Independence Day clashes with Samuel Johnson's birthday, an entry we'd already written, so Peru it is.

~

60ml (2 fl oz) pisco | 30ml (1 fl oz) lime juice
15ml (¾ fl oz) sugar syrup | 1 egg white
Ice | 3 dashes of Angostura bitters

- Shake the first four ingredients in a cocktail shaker, then add ice and shake again. Strain into a rocks glass, over ice or not, depending on your preference. Top with three dashes of Angostura bitters and swirl across the top.

---

29 JULY

COCKTAIL RELEASED (1988)

PINK SQUIRREL –

Cocktail is the finest bartender film ever made, not least because it's the only true bartender film ever made. Purists point out Flanagan is the world's worst technician behind the stick, while his 'Last Barman Poet' poem is cringeworthy and includes a couple of erroneous drinks. The Pink Squirrel does exist, at least, although we don't make many of them.

~

20ml (¾ fl oz) crème de noyaux | 20ml (¾ fl oz) white crème de cacao
40ml (1¼ fl oz) heavy cream | Ice | Grated nutmeg, to garnish

- Shake all the liquid ingredients in a cocktail shaker with ice, then strain into a chilled cocktail glass. Garnish with grated nutmeg.

---

30 JULY

BIRTH OF EMILY BRONTË (1818)

SORBET SOUR – RASPBERRY + RHUBARB, NORTHERN BREWING CO. –

'Out on the wiley, windy moors,' sang Kate Bush in her song 'Wuthering Heights'. Rather fittingly, Emily Brontë, who wrote the book that inspired the song, was something of a Yorkshire gale. Blowing off a fierce wind of change in the world of literature, Brontë eschewed the pre-ordained and male-driven expectation of a woman in Victorian society.

In honour of the writer, and the windy moors she depicts, try some fibrous Yorkshire rhubarb. Used for digestive relief in history's earliest medical records, this vegetable was so revered for inducing bottom burps that by the 17th century it fetched three times the price of opium.

For those less interested in regulating their intestinal transit, though, try rhubarb beer. North Brewing Co. teamed up with ice-cream makers Northern Bloc for this creation, a blend of triple-fruited sour with vegan ice cream, using forced rhubarb from the famed E. Oldroyd & Sons based in Yorkshire's rhubarb triangle.

Today is also the great Kate Bush's birthday. Lovely stuff.

31 JULY

Black Tot Day (1970)

BLACK TOT RUM –

Black Tot Day marks the end of the Rum Ration, a 300-year-old naval perk that saw a measure of rum served up to sailors every day.

Originating in the 17th century, the daily 'tot' was originally safer than water, and, after a late morning cry of 'Up Spirits', the boozy elevenses were gleefully sloshed down seafarers' gullets. The rum bosun (actual job) collected the ration for the mess in rum pails known as a fannies, before salty seamen dipped into said fanny during their daily bouts of gratification.

By the 20th century, the tot was still contractually linked to naval pay, and the ration was 71ml of almost 55 per cent ABV spirit. This is still a reasonable whack, of course, and implies naval officers were getting lively on their morning liveners while controlling nuclear submarines. Sounds like a right laugh, but no one wants something long, hard and full of het-up seamen charging at them. Don't drink and drive, especially nuclear subs.

In 1969 the Admiralty Board (only boring people get board) ruled the ration unsafe, and, on this day in 1970, the last rum was poured while sailors donned black armbands and flung their retired fannies into the sea.

Drink Black Tot rum in salute today: it blends historic rums from the Royal Navy cellars with marvellous modern spirits. Sip and flick a 'V for victory' at all the risk assessors out there.

AUGUST

1 AUGUST

Yorkshire Day

TRANSMISSION IPA, NORTH BREWING –

Nah then, today's the day when Yorkshire, and its flat-capped diaspora, celebrate the historic contributions of 'God's Own County', such as puddings, cat's eyes (see 15 March), MP William Wilberforce (whose Slavery Abolition Act came into force on this very day in 1834), and this 'reet gradely' IPA from the fine folk at North Brewing.

2 AUGUST

Death of Fela Kuti (1987)

ZOMBIE –

Fela Kuti is history's funkiest freedom fighter. The creator of Afrobeat, (a blend of Yoruba rhythms, American funk and pidgin English), he ridiculed Nigeria's military regime with songs like 'Zombie', which mocks military mindlessness, and 'International Thief Thief', a funky condemnation of corporate greed.

As his popularity grew, so did the efforts to silence him. Following the success of 'Zombie', the military repeatedly raided the independent 'Kalakuta Republic', a communal compound housing Kuti's family and friends. It contained a free health clinic, his recording studio and lots of sex and drugs – especially marijuana.

During one bust, officers tried to plant drugs on Kuti. He grabbed the joint, ate it and was arrested. As Police waited for the 'proof' to appear naturally, Kuti hoodwinked officers and deposited it secretly in a communal bucket.

When Kuti finally 'logged-on', three days after his arrest, there was no 'pot' in the pot. A year later, Kuti released a song entitled 'Expensive Shit'– a real favourite with all the lads down the station.

Things got really nasty in 1977, when police set the compound alight and dragged Kuti out of his bed by his genitals, breaking his skull, arm and legs. His 82-year-old mother, meanwhile, was pushed out of a window – suffering injuries she would subsequently die from.

Often giving interviews in nothing but bikini briefs, Kuti was an eccentric Fela. He married 27 women in a single ceremony, employed a rota system to keep them all content, and divorced them all nine years later. 'No man,' he declared, 'has the right to own a women's vagina.'

Having been arrested more than 100 times, beaten and wrongly imprisoned for both currency smuggling and murder, he died from an AIDS-related illness in 1987. But he remains one of Africa's greatest ever musicians. Honour him with a Zombie cocktail.

~

Ice | 50ml (1½ fl oz) white rum | 25ml (¾ fl oz) dark rum
20ml (¾ fl oz) triple sec | 20ml (¾ fl oz) orange juice
20ml (¾ fl oz) lime juice | 15ml (½ fl oz) sugar syrup
15ml (½ fl oz) grenadine | 2.5ml Pernod Absinthe
Grapefruit zest twist and a cherry, to garnish

- Stir in a tall ice-filled glass. Garnish with grapefruit zest and a cherry.

---

3 AUGUST

JESSE OWENS WINS GOLD AT 1936 BERLIN OLYMPICS

GOLD RUSHT –

The 1936 Berlin Olympics gave Adolf Hitler a platform to proudly parade the Third Reich's power, reinforcing the Nazi idea of Aryan racial superiority.

Yet glorification of his regime was undermined right from the opening ceremony, when 25,000 pigeons, ostentatiously released into the 100,000-seater Olympic Stadium and stunned by a deafening cannon shot, emptied their bowels all over the competitors below.

It was not the only 'rain' to fall on Hitler's parade. African-American athlete Jesse Owens won four gold medals – in the 100 metres, long jump, 200 metres, and the 4 × 100-metre relay. 'Floating across the track like water', Owens smashed the world record on the way to the final, leaving Germany's golden boy, Erich Borchmeyer, flailing back in fifth.

While it made a mockery of National Socialist ideology, Hitler refused to acknowledge Owens' achievement. He genuinely believed African-Americans were simply animals with greater physical strength than 'civilised whites' and therefore had no right to participate in the Games.

Owens didn't even know, or care, that Hitler was watching. 'I saw the finish line and knew that 10 seconds would climax the work of eight years,' he said afterwards. 'One mistake could ruin those eight years. So, why worry about Hitler?'

Owens was more hurt by the racial prejudice he encountered back home. After returning from Berlin, where he'd shared his hotel with white teammates, Owens was refused entry to a hotel reception being held in his honour at The Waldorf, New York.

As a Black man, Owens was prohibited from using the main entrance and was unceremoniously bundled up to his own celebration in the hotel's freight elevator. He wasn't even congratulated by President Franklin D. Roosevelt. 'I wasn't invited to shake hands with Hitler,' he said. 'But I wasn't invited to the White House to shake hands with the president, either.'

He'd returned from Hitler's HQ shining with gold, yet still couldn't sit at the front of an American bus. And, when his Olympics money ran out, Owens resorted to running against racehorses as a novelty act. 'What was I supposed to do? You can't eat gold medals.'

Eventually, America recognised his greatness, awarding him the Medal of Freedom in 1976, America's highest civilian distinction.

Today, he is rightly regarded as the greatest sprinter that was ever born. Mark the day of his victory with a Gold Rush cocktail, created at Milk & Honey, New York.

~

50ml (1½ fl oz) bourbon | 20ml (¾ fl oz) lemon juice
20ml (¾ fl oz) honey syrup | Ice

- Shake all the ingredients in a cocktail shaker with ice, then strain into an ice-filled Old Fashioned glass.

---

4 AUGUST

SUPER SATURDAY (2012)

LONDON PRIDE, FULLER'S BREWERY –

At the 2012 London Olympics, Great Britain did something very un-British. It won six gold medals in the greatest day in its 104 year-old Olympic history.

After scooping a couple of golds in rowing in the morning, an amazing afternoon in the Velodrome ensued, with the women's cycling team pedalling to glory. Team GB's efforts peaked in a climax of athletic ecstasy at the Olympic Stadium in front of an incredulous 80,000 crowd.

Mo Farah (10,000 metres), Greg Rutherford (long jump) and Jessica Ennis (heptathlon) delivered three golds in just 26 amazing minutes.

---

5 AUGUST

OYSTER DAY

PORTER –

Back in the 18th century, people popped to the pub for a pint of porter and a few oysters – which, back then, were a working-class snack sitting freely on the bar.

We're not sure when oysters went all posh, but they did. Perhaps it was because of their reputation as an aphrodisiac – Casanova famously scoffed 50 oysters a day, and he got loads of action.

Apparently, the zinc-rich shellfish creates an upsurge in testosterone and improves potency in one's 'gentleman's relish'. Because they look like female genitalia, oysters also increase male sexual desire.

What about women? Well, if they're into other women, then the resemblance to female genitalia applies. Otherwise, the element of danger associated with oyster consumption heightens arousal. Women, eh?

Either way, it's not going to shuck itself is it? So, don't be shy: have an oyster with some porter.

---

6 AUGUST

Jamaican Independence Day

RED STRIPE LAGER –

After more than 500 years of colonial rule, Jamaica claimed its independence from the UK in 1962.

Each year, everyone celebrates in the national colours of green, black and gold, while the main event is the Jamaica Festival in Kingston. It's a colourful carnival of parades, fireworks and the famous Jamaican Marching Band (featuring a man with a triangle, whose job is to stand at the back n'ting).

7 AUGUST

DEATH OF OLIVER HARDY (1957)

BOXCAR –

Notwithstanding the Thinking Drinkers, Stan Laurel and Oliver Hardy remain the greatest comedy duo of all time.

Two friends without a brain cell between them, yet united by a hopeless, endearing optimism, they were both black belts in buffoonery capable of turning the most banal situation into a whirlwind of farce.

Their favourite drink was a White Lady (see 16 June) but we've riffed the recipe here to make a Boxcar, using lime juice, grenadine and Broker's Gin: a classic London Dry that comes with a bowler hat atop each bottle.

~

60ml (2 fl oz) Broker's gin | 25ml (¾ fl oz) lemon juice
25ml (¾ fl oz) triple sec | 5ml (¼ fl oz) sugar syrup
10ml (¼ fl oz) grenadine | 1 egg white | Ice

- Shake all the ingredients in a cocktail shaker with ice, then fine-strain into a sugar-rimmed coupe.

8 AUGUST

ALBERTO SANTOS-DUMONT CRASHES HIS AIRSHIP INTO A PARISIAN HOTEL (1901)

CACHAÇA SPRITZ –

Alberto Santos-Dumont was the first person to fly a personal flying machine.

An eccentric, bon-vivant Brazilian who beguiled Belle Epoque Paris with his aeronautical derring-do, Dumont put the 'high' into high society. During the day, he'd fly to the shops, while, by night, he'd soar down the Champs-Elysees for supper, tying his blimp to a lamppost while he dined.

Flights, however, didn't always go well. On this day in 1901, he circled the Eiffel Tower, lost hydrogen, and hurtled into the Trocadero Hotel. Thankfully, the only injury incurred was to his pride.

Sadly, horrified by how the First World War had mutated his beloved airplane into a killing machine, Dumont took his own life in 1932.

~

Ice | 50ml (1½ fl oz) cachaça
Tonic water | Lime wedge, to garnish

- Fill a glass with ice. Add the cachaça, then top with tonic water and stir. Garnish with a lime wedge.

---

9 AUGUST

CONSTRUCTION BEGINS ON THE TOWER OF PISA (1173)

PISA LIQUEUR (PACKAGED IN A 'LEANING' TOWER-SHAPED BOTTLE) –

When legendary soul singer Edwin Starr repeatedly exclaimed that war was good for absolutely nothing, he obviously hadn't built a 12th-century basilica.

For when the people of Pisa started laying the foundations for their 56-metre (184-feet) belfry in porous clay soil back in 1173, the sheer folly soon became apparent. Luckily for them, war soon kicked off between the Italian states (again) and construction stopped. This gave the tower time to settle in the soil, preventing it from collapse.

When work restarted, it was still leaning. But the Tuscans, much like some cowboys who did a botch job on one of our kitchens in 2011, cracked on anyway, hoping that the problem would simply go away.

By the time they finished in 1372, having stuck some really heavy bells at the top of it (which surely can't have helped), the tower was completed with a 1.4-metre (4.6-foot) lean. Four hundred years later, it had increased to 3.8 metres (12.5 feet), and by 1993, things got wonkier still, with a 5.4-metre (17.7-foot) lean. Unsafe for visitors, the tower was forced to close until 2001.

By 2018, building work had clawed back 4cm of the tilt, and teeth-sucking builders reckon that, with a bit of 'bish-bosh-shoom-shoom-done', the tower will straighten up completely by the year 2300.

10 AUGUST

King Charles II and John Flamsteed lay the foundation stone of the Royal Observatory in Greenwich, London (1675)

LONDON PALE ALE, MEANTIME BREWING –

When King Charles floated the idea of a Royal Observatory in Greenwich, a lot of people were extremely sceptical. But once they'd peered down its huge telescope, they could really see where he was coming from.

The Observatory was instrumental in the adoption of the Greenwich meridian as the starting point for international time zones in 1884. Greenwich is also home to Meantime Brewing, London's first 'craft' brewery, set up in 1999 by the pioneering Alastair Hook. Its Pale Ale is lovely, made using hops sourced from the Kent countryside (which is a dangerous thing to say after a few pints).

11 AUGUST

A GREENLAND SHARK DECLARED OLDEST VERTEBRATE ANIMAL IN THE WORLD BY INTERNATIONAL TEAM OF SCIENTISTS (2016)

EVEN SHARKS NEED WATER IPA, VERDANT BREWING –

The Greenland shark is an apex killer, sitting at the top of an Arctic food chain featuring other sharks, seals, fish, whole deer and, if they're thick enough to be paddling about nearby, humans.

Given its notorious reputation as the ultimate Arctic assassin, one wouldn't advise anyone to approach it and start dicking around with one of its eyes. But that's exactly what some Speedo-wearing foolhardy scientists did in 2016.

After bravely scraping carbon-dating proteins from the lenses of the Greenland shark's eye, they discovered it was born in the early 1600s. This means it was wearing nappies when the Pilgrim-packed ship *Mayflower* set off for America (see 6 September), and hit sexual maturity, aged 150, when Captain Cook discovered New Zealand and Australia.

12 AUGUST

WORLD ELEPHANT DAY

DELIRIUM TREMENS –

A pink elephant adorns the label of this strong (8.5 per cent) Belgian Pale Ale. Enjoy one while we tell you stuff about elephants that you probably don't know. Or even need to know.

There are two types of elephant in the world – Asian and African

– but they're too biologically different to breed.

An elephant only has two knees, on its back legs, because the front legs are arms – the bendy joint being elbows.

An elephant can hear better when it has one foot off the ground.

The huge, dangling, pendulous appendage that is the elephant's trunk has more in common with the human tongue than any other human organ (including the penis.) Humans, however, can't lift a 150kg tree trunk with their tongues (and only some of us can do that with our penis).

An elephant's penis was famously used to fashion a golf bag for King Edward VII.

Elephants are scared of two things: bees and humans (the latter presumably because they fashion their dongs into golf bags).

---

13 AUGUST

WAGNER'S *RING* FIRST PERFORMED

8 BALL RYE IPA, BEAVERTOWN BREWERY –

Opera, famously, 'isn't over until the fat lady sings.' Which is confusing for uninitiated opera-goers, as singing fat ladies tend to feature throughout.

The grand operas of the 19th century famously reached their zenith with Richard Wagner's *Ring* cycle: four operas about Nordic Gods, malign dwarfs and mortal heroes, spread over 15 hours in what Wagner called 'music drama.' We strongly suggest you have a wee before it starts.

It's said that the famous 'fat lady' phrase was coined by the opera-enamoured Chicago gangster Al Capone. When one of his goons stood up to leave after the end of the first aria, Capone grabbed him by his coat and growled: 'Siddown … it ain't over until the fat lady sings.'

Another theory, however, has nothing to do with opera, instead proposing that the expression is merely a misquote relating to the game of pool. 'It's not over until the fat lady *sinks*' refers to the name for the black eight ball (the 'fat lady'), which is always the last to be potted.

So we're suggesting the 8 Ball from North London's Beavertown Brewery: a mash up of spicy rye and zesty hops first created using old pool balls to weigh down the hop sack during the brewing process.

14 AUGUST

COLOGNE CATHEDRAL COMPLETED (1880)

KÖLSCH –

Cologne in Germany is not the prettiest of places, even if you're wearing beer goggles.

What once stood regal on the Rhine was reduced to rubble in the Second World War by 34,711 tonnes of bombs, which destroyed more than 3,000 buildings – but not Cologne's magnificent cathedral. Having taken more than 632 years to build, it point-blank refused to buckle under the bombardment and, despite taking more than a dozen direct hits, survived: a perpendicular emblem of faith, resplendent among the city's ruins.

Equally resplendent is Cologne's beautiful beer. *Kölsch* (brewed like an ale, yet matured like a lager in cold conditions) is served in small, session-friendly 20cl glasses. This keeps the beer fresh and encourages the companionable buying of rounds.

15 AUGUST

INDIAN INDEPENDENCE DAY

RAJ IPA, TRYST BREWERY –

Impassioned by the decades-long, non-violent campaign against British rule embodied by flip-flopped pacifist Mahatma Gandhi, India gained independence in the summer of 1947.

Every year, at the Red Fort in Delhi, the Indian prime minister raises the national flag and issues a 21-gun salute in remembrance of those who led the Indian independence movement. We're raising an India Pale Ale, a beer style initially brewed to withstand the long sea journey from Britain to Bombay.

16 AUGUST

BIRTH OF MADONNA (1958)

LANDLORD PALE ALE, TIMOTHY TAYLOR –

In 2003, Madonna declared her love of Landlord, the award-wining ale from the quiet, cloistered Timothy Taylor Brewery in Keighley, Yorkshire.

She'd been converted by her then-husband, film director Guy 'Toff Guy' Ritchie, at the height of her tweed flat-cap phase. Predictably, sales of the Yorkshire bitter rose 'Like a Virgin' ogling her 'Justify My Love' video, and the brewery's vans displayed her claim that Landlord was 'the Champagne of Ales'.

17 AUGUST

DAVID BECKHAM SCORES ICONIC GOAL FROM THE HALFWAY LINE (1996)

HAIG CLUB WHISKY –

In the last minute of the first match of the 1996–1997 season, Manchester United's David Beckham received the ball on the right side of his own half. Looking up and spotting the Wimbledon goalkeeper Neil Sullivan off his line, the 21-year-old swept his right foot through the ball, sending it soaring over a scrambling Sullivan and into the back of the net – from 55 yards.

'When my foot struck that ball,' 'wrote' Beckham in his autobiography, 'it kicked open the door to the rest of my life.'

That remarkable strike, followed by the midfielder's arms-out, Messiah-like pose to the cameras, announced Beckham's arrival on the global stage – not just as a footballer but, ultimately, as a cultural icon.

'I couldn't have known it then,' he added in his book, 'but that moment was the start of it all: the attention, the press coverage, the fame.'

After hanging up his boots in 2013, Beckham launched his own whisky. Haig Club, packaged in a stylish blue bottle, is a single-grain whisky from the Cameronbridge distillery, which was opened in 1824 by John Haig, a pioneer in column-still grain whisky.

The grain is nearly all wheat, with 10 per cent malted barley, and is aged in ex-bourbon barrels for around seven years: Beckham's number. Among single-malt snobs, it remains as unsung as Posh Spice's back-catalogue – but let's face it, it's not for them. It's deliberately directed at whisky drinkers who are happy to mix it with Coke and, ahem, blend it like Beckham.

---

18 AUGUST

FRENCH ASTRONOMER PIERRE JANSSEN DISCOVERS HELIUM IN SOLAR SPECTRUM DURING ECLIPSE (1868)

ECLIPSE COCKTAIL –

Solar eclipses have scared the shit out of people for centuries. The Pomo, an indigenous tribe from California, believed that a bear was taking huge bites out of the sun. Even back then, that seems unlikely. Inuits, meanwhile, reckoned the dual rays caused sickness – so they did the obvious thing and turned their cutlery the other way round to reflect the bad vibes away from them. Clever.

Animals act oddly, too. In 1932, professors noticed that dogs got spooked and birds returned to their coops. Interestingly, monkeys carried on as normal – swinging in tyres, flinging poo, that kind of thing.

In 1868, Pierre Jules César Janssen, a bon-vivant boffin studying an eclipse in India, spotted some bright yellow lines in the second outer layer of the sun. He deduced that the cause of the colour was an undiscovered element. Norman Lockyer, a British astronomer, had also spotted the same colour, which he named 'Helium' after Helios, the Greek god of the sun.

Sadly, following his helium breakthrough, Janssen struggled to get further research published because nobody could take him seriously with that silly voice. However, he did write a fascinating book on the subject (we *literally* couldn't put it down).

~

50ml (1½ fl oz) Añejo tequila | 35ml (1 fl oz) Aperol
35ml (1 fl oz) Heering Cherry liqueur | 35ml (1 fl oz) lemon juice
Ice | Mezcal, for rinsing | Lemon zest twist, to garnish

- Shake the first four ingredients in a cocktail shaker with ice. Strain into an ice-filled Old Fashioned glass that's been given a swirl with mezcal (which is then discarded). Garnish with a lemon twist.

---

19 AUGUST

BIRTH OF COCO CHANEL (1883)

POMME POMME COCKTAIL –

French fashion designer Gabrielle Bonheur 'Coco' Chanel created some of the world's most famous clobber, including pea coats, bell-bottoms, the 'Chanel' suit and, of course, the 'Little Black Dress'.

Prepare yourself a Pomme Pomme: a twist on the classic champagne cocktail. It combines Calvados (the brandy from Normandy, home to Chanel's first shop in Deauville) and French bubbly, her favourite drink. Her logo also looks like two apples. Sort of.

~

Sugar cube | 2 dashes of Angostura Bitters
25ml (¾ fl oz) Calvados | Champagne, to top up

- Soak a sugar cube with a couple of dashes of Angostura, then place in a champagne flute. Cover with Calvados, give it a stir and top up with chilled champagne.

---

20 AUGUST

World Mosquito Day

MOSQUITO PUNCH –

Mosquitos. Quite why there's a day celebrating the blood-sucking bell-ends is beyond us, but today's alternative option was a Russian who discovered Alaska – and that required *research.*

So, instead, some mosquito facts. Mosquitos are the world's deadliest animal, responsible for killing more than half the people that have ever lived. Ever. By spreading fatal diseases, like malaria, dengue fever and Zika, their body count hits 1 million every year.

Only female mosquitos bite, and only the silent ones can give you malaria. A buzzing mosquito will still give you all the other lurgies. And keep you awake. Given the choice, mosquitoes bite the ankle – often drinking three times their own body weight in blood – and if you've just eaten a banana, they find you extra tasty.

So why don't we just kill them all? Well, apparently they're prime pollinators, so the ecosystem would collapse. However, with an average lifespan of just two months, it doesn't really matter if you twat one with a newspaper.

~

25ml (¾ fl oz) white rum | 5ml (¼ fl oz) ginger cordial
5ml (¼ fl oz) lemongrass cordial | 5ml (¼ fl oz) lemon juice
Ice | Soda water, to top up | Cucumber slice, to garnish

- Pour the rum, cordials and lemon juice into a glass filled with ice. Top up with soda water, then stir. Garnish with a cucumber slice.

21 AUGUST

Mona Lisa stolen (1911)

MONA LISA COCKTAIL –

Few outside the art world had set eyes on the iconic enigmatic smile of the Mona Lisa before it was nicked from the Louvre.

The world's most famous painting was stolen by three Italian handymen, who simply detached the canvas from its frame, threw a towel over it, walked out of the museum and caught a train to Italy.

News of the missing masterpiece quickly spread, and the iconic image of Mona Lisa looking like she'd cycled down a cobbled street was splashed over front pages of newspapers all over the world.

Amid growing hostilities ahead of the First World War, some suspected zee Germans of stealing it. Famously, Pablo Picasso was also brought in for questioning before being quickly released.

As ever, it was an inside job. However, the chief thief, Vincenzo Perugia, was unable to flog Leonardo Da Vinci's masterpiece – its instant overnight international fame made it too hot to hawk.

After trying to flog it to a Tuscan dealer, who told the police, Perugia and his pals were arrested and given eight months in prison. Given the painting's now worth an estimated $850 million, it was almost worth it. Mark the occasion with a Mona Lisa cocktail.

~

40ml (1½ fl oz) Chartreuse | 70ml (2½ fl oz) orange juice
2 dashes of Angostura bitters | Ice
Tonic water, to top up | Orange slice, to garnish

- Shake the Chartreuse, orange juice and bitters in a cocktail shaker with ice. Strain into a highball glass packed with ice and top up with tonic water. Garnish with an orange slice.

22 AUGUST

Richard III killed at Battle of Bosworth (1485)

SKULL SPLITTER, ORKNEY BREWERY –

More than 520 years after he was bopped on the head by a Welshman and killed at the Battle of Bosworth, Richard III was discovered underneath a Leicester car park.

Not content with inspiring the expression 'Busting for a Richard', he is the most villainous of all English monarchs, a hunchbacked sadist who killed babies, slaughtered wives and sported a 'bowl' haircut that won't have been fashionable even back then.

The last king to die on the battlefield, Richard III was reburied in Leicester in 2015. This miraculously coincided with an incredible upturn of fortunes for the city's football team, who, against all odds, won the Premier League Championship. The name of Leicester City's stadium? King Power. Coincidence? Yes.

23 AUGUST

Death of Rudolph Valentino (1926)

BLOOD & SAND –

Rudolph Valentino was damn sexy. The erogenous, androgenous Italian lit up the silent screen throughout the first half of the twenties, and was everything an American man wanted to be, but probably couldn't.

The 'Latin Lover' danced the tango, the cakewalk and the bunny bug; he rode horses, he could sing, write poetry and, while known to weep now and again, he was also a talented boxer who could smasha-your-face in. Chicks dig that shit.

When he died from appendicitis complications, aged just 31, thousands of tearful mourners lined the streets of New York, and some fans, distraught at his death, took their own lives.

One of his most famous films, *Blood & Sand* (about a bullfighter with two birds on the go), is also the name of a well-known whisky drink from Harry Craddock's 1930 *Savoy Cocktail Book.*

~

20ml (¾ fl oz) Scotch whisky | 20ml (¾ fl oz) Heering Cherry liqueur
20ml (¾ fl oz) sweet vermouth | 25ml (¾ fl oz) orange juice
Ice | Flamed orange peel, to garnish

- Hard shake all the ingredients (except the garnish) in a cocktail shaker with ice. Strain into a chilled cocktail glass. Garnish with a flamed orange peel.

---

24 AUGUST

Pliny the Elder Dies (79 CE)

PLINY THE ELDER DOUBLE INDIA PALE ALE, RUSSIAN RIVER BREWING –

Pliny the Elder was an esteemed Roman author, naturalist and Greek philosopher. He'd make a shit doctor, though. To cure epilepsy, he suggested eating the heart of a black jackass, outside, on the second day of the new moon. Alternatively, you could munch some lightly poached bear testes or a dried camel brain with honey, or drink fresh gladiator's blood.

To treat incontinence, try touching the tips of your genitals with papyrus. If that doesn't work (which it won't) drink a glass of wine mixed with the ash of a burned pig's penis. Or urinate in your neighbour's dog's bed.

A cream made from pig lard and the rust of a chariot wheel cures haemorrhoids, as does swan fat and the urine of a female goat. Headaches? Tie some fox genitals to your head. As for hangovers, Pliny prescribes some raw owl's eggs or fried canary.

Some beer historians claim Pliny was the first person to write about hops in his *Naturalis Historia*, a huge survey of contemporary human knowledge. He wasn't. The 16th-century Swedish botanist Carl Von Linné actually gave the hop its botanical name – but Linné is nowhere near as funny.

Even the way Pliny the Elder died is amusing. When Mount Vesuvius erupted and destroyed Pompeii, Pliny went to investigate the eruption, wearing a pillow tied to his head as protection.

It didn't work. He died.

Pliny the Elder is a highly-hopped, 8 per cent beer from Russian River Brewing in Santa Rosa, deep in Californian wine country. The ultimate double IPA, brewed with five hops, still commands an enormous cult following and is one of the world's great beers.

---

25 AUGUST

WHISKEY SOUR DAY

WHISKEY SOUR –

Want to look well cool in a cocktail bar? Then order yourself a Whiskey Sour. It suits all seasons and is a concrete classic. The egg is optional, but bulks out the drink's body.

~

50ml (1½ fl oz) bourbon whiskey | 25ml (¾ fl oz) lemon juice
25ml (¾ fl oz) sugar syrup | ½ egg white | Ice

- Shake all the ingredients in a cocktail shaker with ice. Strain over ice into a rocks glass.

26 AUGUST

William Perkins invents a new dye ... by accident (1856)

AMETHYST AVIATION –

Before British chemist William Henry Perkin accidentally created the world's first synthetic organic dye, all colours came from natural sources, such as leaves or flowers.

His unintended creation was purple-ish in hue. He named it mauveine, and the dye transformed the world of fashion, making hitherto wildly expensive violet garments immediately affordable for everyone. Mauveine became absurdly popular – especially after Queen Victoria rocked a gown dyed with his discovery at the 1862 Royal Exhibition. However, it was soon usurped by other synthetic dyes, and now not even pint-sized popstar Prince could make it fashionable again.

Make yourself a mauve-coloured Amethyst Aviation using violet-hued gin. The ancient Greeks reckoned Amethyst, a violet version of quartz, prevented drunkenness. The term derives from *amethustos*, meaning 'not intoxicated'.

~

40ml (1¼ fl oz) Boë Violet Gin | 25ml (¾ fl oz) fresh lemon juice
10ml (½ fl oz) maraschino liqueur | Ice
Fentiman's Grapefruit Tonic, to top up
Fresh pink grapefruit slice, to garnish

- Shake the gin, lemon juice and maraschino liqueur in an ice-filled cocktail shaker, then strain into a tall glass filled with ice. Top up with grapefruit tonic and garnish with a grapefruit slice.

27 AUGUST

Shortest Ever War (1896)

SHOT OF RUM –

In 1896, the British told the sultans of Zanzibar to stop mucking about. When they didn't, British Navy ships opened a can of whuppass on the poor sods and won the war in 38 minutes.

28 AUGUST

St Augustine's Day

MAXIMATOR, AUGUSTINER-BRÄU –

Before famously 'finding' God in his thirties, St Augustine was a rascal – hanging with some cool cats in Algeria, fathering a son, wooing womenfolk and routinely getting hammered.

'Grant me chastity and continence,' he famously asked the Almighty, before setting off down the path of salvation, 'but not yet.'

After a religious epiphany in Milan, he kickstarted England's conversion to Christianity, and, fittingly, God made him the patron saint of brewers. So have a glass of Maximator, a colossal copper-coloured Doppelbock from Augustiner-Bräu, Munich's oldest brewery, which was built by 14th-century hermits.

It's 7.5 per cent, so if you drink too much of this, you won't be able find your keys, let alone God.

29 AUGUST

BIRTH OF HARRY CRADDOCK (1875)

SAVOY SPECIAL NO.2 -

Schooled in the art of bartending at some acutely dapper drinking establishments in New York and Chicago, Harry Craddock famously served the last legal drink before Prohibition began in the US, then headed back to Blighty.

After several years working under the legendary Ada Coleman at The Savoy, Craddock became head bartender at the hotel's iconic American bar in 1926. Coleman had retired, and The Savoy felt Craddock's American accent would appeal to its huge American clientele.

In 1930, Craddock wrote *The Savoy Cocktail Book*, a curation of more than 1,000 recipes encountered during more than three decades behind some of the world's finest bars. An iconic cocktail tome, it remains on the shelf of every serious bartender today.

His legacy also remains somewhere within the walls of the Savoy's American Bar, where, in 1927, he famously hid a shaker containing his favourite White Lady cocktail (see 16 June). It's never been found.

Honour his memory with a Savoy Special No. 2.

~

40ml (1¼ fl oz) gin | 20ml (¾ fl oz) dry vermouth
2 dashes of Dubonnet Rouge | Ice
Orange zest twist, to garnish

- Shake the gin, vermouth and Dubonnet in a cocktail shaker with ice. Strain into a chilled cocktail glass and squeeze a twist of orange peel on top.

30 AUGUST

'HEY JUDE' BY THE BEATLES RELEASED IN THE UK (1968)

HACKER PSCHORR ANNO 1417 MUNICH KELLER BIER –

In Germany, where McCartney was arrested for setting fire to a condom in 1960, 'Hey Jude' is a classic crowd-pleaser among stein-swinging drinkers at the Munich Oktoberfest.

31 AUGUST

BIRTH OF VAN MORRISON (1945)

MOONDANCE BEST BITTER, TRIPLE FFF BREWERY –

Since peaking early at the age of 23 with *Astral Weeks*, one of the greatest albums of all time, Van Morrison tends to polarise opinion. Still, it's a wonderful night for a Moondance.

SEPTEMBER

1 SEPTEMBER

LAST PASSENGER PIGEON DIES (1914)

PINK PIGEON SPICED RUM –

If you live on Pigeon Street, one of the pigeons you won't meet is the passenger pigeon, because it's extinct. The last of its kind sadly died in captivity in Cincinnati Zoo on this day in 1914.

There are lots of other species of pigeon, of course, so why get in a flap about the loss of this one? Well, its demise was indicative of the impact humans continue to have on natural wildlife. In the early 1800s, as the Europeans expanded westward through the New World, billions of passenger pigeons would flock in such numbers they could eclipse the sun. And yet 100 years later, they were hunted to extinction.

It's worth adding the passenger pigeon was a bit of a pest – at their peak population, their poop was a genuine pollutant for people. Lot of 'P's there, and here's a couple more in Pink Pigeon rum, which is produced in Mauritius and is named after the rare species of the same name, which recently hopped from the 'critically endangered' to 'endangered' list. This sweet and easy drinking rum is spiced with botanicals including vanilla and nutmeg.

2 SEPTEMBER

GREAT FIRE OF LONDON STARTS (1666)

CRAFT LAGER, TOAST ALE -

We weighed up whether we could rise to the challenge and sprinkle some half-baked puns through this one. This is what we get paid the big dough for, and such quips might have been the icing on the cake. But we know we knead to grow up, because people really were overcooked in the Great Fire of London.

The fire started in a bakery in Pudding Lane in 1666, destroying most of London, and while only six people died, 70,000 homes were vapourised.

Not necessarily an event to toast today, then, but if you are having a drink, try not to get baked while enjoying Toast Ale. As well as producing award-winning beer, the folks at Toast brew from surplus fresh bread that would otherwise go to waste.

3 SEPTEMBER

SWEDEN BEGINS DRIVING ON RIGHT-HAND SIDE OF ROAD (1967)

MACKMYRA WHISKY -

We fear change – that's why we've worn the same underpants for the last 20 years. So, we sympathise with the Swedish, who whoopsied their own perished pants when the government told them to switch from driving on the left to the right in 1967.

The change would bring the Swedes into line with neighbours, but perhaps more relevant was the fact that nearly all of them were driving left-hand vehicles anyway. Even so, the motorists were worried, convinced it would generate all the inconvenience, irritation and injury of minimalist flatpack furniture construction. They even dubbed the event 'Dagen H', and while this sounds like a sinister science experiment, it simply stands for 'H Day' – the 'H' is for '*Högertrafikomläggningen*', or 'the right-hand traffic reorganisation', which is considerably less thrilling and arguably doesn't warrant this much coverage in our book.

Despite their concerns, the change worked, and the triumph possibly paved the way for Sweden's avant-garde Mackmyra to take whisky in a new direction. This pioneering distillery is fearless in its attempts to buck traditions. Their Svensk Ek whisky is matured in barrels made from oak trees planted on the island of Visingsö centuries ago, while their Grönt Te, a collaboration with Japanese tea specialists Yuko Ono Sthlm, sees single malt finished with tea leaves. Sounds whacky, but all their whisky tastes fantastic – and, since you can't drink and drive, it won't cause traffic accidents.

---

4 SEPTEMBER

World Sexual Health Day

PAINKILLER COCKTAIL –

Are you suffering from some chaffing under 'Kojak's rollneck'? Are you having some 'gearbox trouble'? Are there bats flying around your bell tower? Then, oh, think twice, or it's another day of dealing with your pubic lice, and get yourself down the clinic for a check-up. For today is 'World Sexual Health Day', designed to encourage more people to get regular sexual health checks (also known colloquially as 'VD Day').

For anyone who has already had a rigorous once-over, try a Painkiller cocktail. While useless at curing 'underpants ouch', it's potent enough to instil the courage to deliver the news to everyone you've slept with in the last month or so.

Don't forget to add a cocktail umbrella – a painful reminder of why it's always sensible to wear protection.

~

60ml (2 fl oz) Pusser's Rum | 120ml (4 fl oz) pineapple juice
30ml (1 fl oz) orange juice | 2 tablespoons cream of coconut
Ice | Freshly grated nutmeg
Orange slice and a cherry, to garnish

- Shake the first four ingredients in a cocktail shaker with ice and strain into a highball glass or goblet filled with ice. Grate fresh nutmeg on top and garnish with an orange slice and cherry.

---

5 SEPTEMBER

World Beard Day

BLACKBEARD COCKTAIL –

Russian Emperor Peter I put a tax on beards and demanded lawbreakers endure a public shaving. Henry VIII also enforced a beard tax, even though he had facial hair himself. And his daughter, Elizabeth I, who we assume didn't have a beard, did the same. So, why so anti-beard, folks?

Perhaps these rulers felt beards were dirty – which is a fair point. Indeed, scientists concluded that beards harbour more germs than dogs carry in their fur – and dogs don't wipe their arses when they take a shit.

If you do like beards, though, then the most famous example comes in the form of pirate Edward Teach, otherwise known as Blackbeard (see 22 November). Since we didn't use this drink there, and there aren't many cocktails with 'beard' in their name, we'll put it here.

~

40ml (1¼ fl oz) spiced rum | 140ml (4¾ fl oz) Coca-Cola
60ml (2 fl oz) Guinness

- Pour into a pint glass in order and serve.

6 SEPTEMBER

MAYFLOWER SETS SAIL

PLYMOUTH GIN –

After several false starts and months of mistakes, boat changes and repairs, the Pilgrim passengers of the *Mayflower* finally sail to the New World, and into the history books. But while they are drenched in celebratory sea spray, there was a fair amount of filthy flotsam on their faces – and, indeed, on their hands.

For a start, no one liked them. Stubbornly committed to strict religious standards, their lifestyle jarred with that of the groovy, care-free Church of England, forcing the Pilgrims to escape.

And while they talked a good game about drunkenness being the devil's work, the *Mayflower* was essentially a floating pub. Designed specifically to transport wine, it was packed with 10,000 gallons of beer, 120 casks of malt for brewing, and 12 gallons of Dutch gin.

Ungodly as it was, the booze came to their aid on more than one occasion. When a storm damaged the ship, it was a screw from their cider press that restored the mast, and when water proved unsafe to drink, beer became an essential source of hydration.

Meanwhile, this glorified booze cruise didn't even navigate to the correct destination. The Pilgrims eventually tossed their tow-rope on to Plymouth Rock, several hundred miles north of their intended target in Virginia, simply because they'd run out of that beer.

But if the ship's navigation was off, it had nothing on the Pilgrims' moral compass. Once they got themselves nice and settled, they slaughtered or enslaved the natives. Thank goodness they preserved those strict religious principles.

So, ignore the Pilgrims and focus on Plymouth, because it's home to the excellent Plymouth Gin Distillery. The oldest working distillery in England, they have been making Plymouth Gin here, according to the original recipe, since 1793.

7 SEPTEMBER

National Salami Day

VIAEMILIA, BIRRIFICIO DEL DUCATO –

We don't want to flog the ubiquitous sausage-and-beer pairing, but lager and salami really does make for a slap-up meal. Take Birrificio Del Ducato's Viaemilia paired with *capicola*. Birrificio's crisp but lightly honeyed and floral award-winning *kellerbier* is delish, carefully created each year when brewers travel to Tettnang in Germany to make their selections from the region's famous and floral noble hops. Meanwhile, the *capicola* (or 'gabagool', according to Tony Soprano) is its ideal partner: the whole-muscle *salume* is seasoned with flavours like wine, garlic and paprika.

8 SEPTEMBER

World Literacy Day

ABC COCKTAIL –

What a marvellous choice you've made by reading this particular book on this special day. UNESCO will be proud.

~

15ml (½ fl oz) Amaretto | 15ml (½ fl oz) Baileys
15ml (½ fl oz) Cognac

- Chill the ingredients, then layer each, in order, in a shot glass.

9 SEPTEMBER

RELEASE OF *A FEW GOOD MEN* (1992)

REGAL ROGUE VERMOUTH –

In this cracking 1992 courtroom drama, Tom Cruise plays Lieutenant Kaffe. During a scene of high-judicial jeopardy, he suggests judge and jury enjoy a wormwood-based fortified wine, to which Jack Nicholson's Colonel Jessop angrily responds: '*Vermouth*? You can't handle vermouth!'

10 SEPTEMBER

BIRTH OF MARIE LAVEAU (1801)

CLAIRIN (HAITIAN RUM) –

Marie Laveau was a 19th-century Louisiana Creole practitioner of Voodoo, and so proficient was she in that particular caper, she was called the 'Queen of Voodoo'. She was also a herbalist, a midwife, and a woman of colour who challenged the injustices of racism and sexism, generally trailblazing her way through life.

While her public rituals were reportedly dramatic, some involving serpents and the blood of sacrificed chickens, much of this drama was a tourist stunt to raise money for the poor. Away from these haunting spectacles, Laveau provided support and empowerment for women.

The theatrical aspect of Voodoo has been sensationalised in popular culture, but at its core, the faith is no more outlandish than Christianity. And, crucially for us, rum plays a pivotal role in some rituals.

Among the statues and idols honouring Laveau is one in which she has her snake, Zombi, around her neck as she draws out Baron Samedi. Samedi is the Voodoo guardian of the cemetery, a dapper ghost who steps into the land of the living donning top hat and tails. He's called on for protection and, when summoned, is usually offered treats. These need to include rum, and Samedi's preference is for raw *kleren* (rum) steeped in 21 hot peppers, which renders it so spicy that it causes extreme discomfort the following morning.

Kleren, or clairin, is a stunning agricole rum, distilled from sugar cane juice rather than molasses, popular in remote regions of Haiti. You can try clairin rums selected from four communes, including Cavaillon, Barraderes, Pignon and St Michel de l'Attalaye. Clairin Sajous is a great place to start: double-distilled at the Chelo distillery in the village of St Michel de l'Attalaye, it's a pure, fruity and herbaceous spirit that the Baron would approve of.

11 SEPTEMBER

HENRY HUDSON DISCOVERS MANHATTAN

MANHATTAN COCKTAIL –

Legend has it that when the Europeans arrived on Manhattan Island in 1609, they discovered the native Lenape people weren't drinking. They were living in wigwams and tepees, they were just two tents. Dyageddit? 'Two tents'. As in, 'too tense' …

The Europeans immediately started brewing and distilling, opening bars, drinking too much and generally making tits of themselves, so the Lenape named the settlement 'Manna-hatta', or 'high land'. This was interpreted in 19th-century book *The Englishman's Guide-book to the United States and Canada* as 'the land of the high or drunk people'.

We've been back and forth to Manhattan over the years, cementing these traditional European drinking values. During one trip with Woodford Reserve, we sampled and judged Manhattans in a collection of the best cocktail bars. The adventure was cunningly named: 'Manhattans in Manhattan'.

Woodford is a wise choice if you plan to stir your own: it contributes to an exceptional Manhattan due to the bold grain and wood, along with its blend of sweet aromatics, spice, and fruit and floral notes.

~

60ml (2 fl oz) Woodford Reserve Straight Bourbon
25ml (¾ fl oz) sweet vermouth | 3 dashes of Angostura bitters
Ice | Cherry, to garnish

- Pour the bourbon, vermouth and bitters into a mixing glass. Add ice, stir gently for 10–15 seconds, then strain into a cocktail glass. Garnish with a cherry.

---

12 SEPTEMBER

FIRST BROADCAST OF *THE SMURFS*

LINDEMANS APPLE LAMBIC –

Some prats have opined that the Smurfs' Phrygian hats represented Ku Klux Klan hoods. Others argued that, since they share all their possessions, the Smurfs must be communists. Both views are, of course, utter bullshine, and neither are supported by the Belgian team who created them. Safe to say, *The Smurfs* is simply a funny little animated show for children and, weirdly enough, not a political parable in any way.

The Smurfs live in the forest and their height is measured in apples – they are three apples tall – hence our suggestion of a traditional Lindemans lambic, which blends apple juice with a wild-yeasted, spontaneously fermented Belgian beer.

---

13 SEPTEMBER

ROALD DAHL DAY

CHÂTEAU DE BEAUREGARD –

An enthusiastic collector of wine, Dahl kept Mouton Rothschild, Lafleur, Léoville-Las-Cases, Pichon-Longueville, Léoville-Barton, Canon, Angélus and Beauregard in his cellar. And, for ease of delivery, he even installed a chute. The chute reminds us of *Charlie and the Chocolate Factory*, when that fatso Augustus Gloop is sucked out of the chocolate river. And Violet Beauregard is actually named after Bordeaux's Château Beauregard, one of the largest Pomerol estates and producer of prestigious wines like Cheval Blanc and Pétrus.

14 SEPTEMBER

Birth of Bong Joon-Ho (1969)

JINRO SOJU –

The South Korean director Bong Joon-Ho deservedly earned an Oscar for his fine film *Parasite*, which is much better than *Snowpiercer*. Soju is a Korean spirit made by distilling rice, grains and potatoes, much like vodka, but with varying strengths from 20 per cent all the way to 50 per cent ABV.

15 SEPTEMBER

Guatemalan Independence Day

BOTRAN SOLERA 1893 –

'*Feliz día de la independencia*,' they might be saying in Guatemala today. Although, since the day marks the country's freedom from Spanish rule, they might also say, '*Ki'imak k'iin u le je'ela*,' which is a little Yucatan Mayan we picked up. Off the internet. There are other ways they might say it, too: there are 25 Guatemalan languages.

Enjoy some Botran rum while you weigh all that up. It's distilled from virgin sugar-cane honey, and the Solera 1893 uses the solera system, maturing the spirit at high altitude in casks previously used for sherry, port and bourbon.

Ki'. (Tasty.)

16 SEPTEMBER

ARNOLD SCHWARZENEGGER BECOMES A US CITIZEN (1983)

PROTEIN SHAKE –

In an interview with *Men's Health*, Arnie predictably recommends a protein shake, but earns genuine *Last Action Hero* status when claiming he terminates the goodness by adding schnapps or tequila to this drink.

~

20ml (¾ fl oz) schnapps or tequila | 250ml (8½ fl oz) almond milk
60ml (2 fl oz) cherry juice | 1 banana | 1 scoop of protein powder
1 raw egg (including shell, because Arnie is well hard)

- Blend all the ingredients in a blender. Arnie drinks his from the blender container.

17 SEPTEMBER

NORMAN BUCKLEY BREAKS THE WORLD WATER SPEED RECORD (1956)

LAKES DISTILLERY WHISKY –

In 1956, Norman Buckley earned a front-page splash when he broke the one-hour world water speed record on Lake Windermere. Water way to have a good time.

Had the Lakes Distillery been operating, Norman might have dipped into their whisky to celebrate. Opened in 2014 and built in a 160-year-old farmstead on the banks of the River Derwent, the folks here have been making waves with an English single malt that really opens up if you add a drop of water.

18 SEPTEMBER

BIRTH OF SAMUEL JOHNSON (1709)

ANTY GIN, CAMBRIDGE DISTILLERY –

Aardvark. Great word, eh? Of Dutch origin, but still a worthy opening to the English dictionary. Well, actually, if we're to be 'a is for accurate', the first 'word' would be 'a' – the *Oxford English Dictionary* separates 'a' into 33 definitions.

But aardvark is up there. It means 'earth pig', apparently. Did you know they swallow food whole and flex their stomach muscles to chew it? Their long snouts have 10 turbinate bones that funnel air through their noses and give them the greatest sense of smell in the animal kingdom. They also eat 'a is for ants', which brings us seamlessly on to Anty Gin.

Anty Gin takes its name from the ants used in the distillation, which lend the gin a distinctive citrus note, owing to the fact that ants spray formic acid as a method of defence. So, while it sounds a bit gimmicky, and possibly offensive if you like ants, the ingredient has a purpose.

But we digress, because it's Samuel Johnson's birthday. Johnson published his own dictionary in 1775, before the aardvark was even known about. So, what was *his* first word? Well, we can tell you. It was 'abacke, *adv*. backwards'.

Were you taken 'abacke' by that? Perhaps not. But as we read on, we see that, after 'abacke', he had 'abactor', to describe those who drive away or steal cattle in herds. Then 'abacus', a counting table. There are loads more 'a' words to go through, but we should wrap this up before our publisher susses our attempts at hitting the contractual word count by bulking out entries with spurious content and superfluous words – even though it was never easier to do than in the case of Johnson and his dictionary.

That said, 'b' is also interesting. Johnson was a tea fan, but occasionally sipped 'b is for brandy'. However, he leads with 'b is for baa', as in the sound of a sheep, which is quite funny. And next he offers 'b is for babble' which means to talk, er, idly ... Oh.

19 SEPTEMBER

First Glastonbury Festival (1970)

SOULSHAKERS' VOODOO PUNCH, AKA 'THE GLASTONBURY ZOMBIE' –

There are few things more tedious than reading about someone else's 'Glasto' experiences. And that truth saves us a lot of time writing something here.

~

1 bottle white rum | 2 bottles light aged rum
1 bottle añejo rum | 1 bottle demerara rum
½ bottle overproof rum | 1 bottle Cognac
1 bottle Orange Curaçao | 1 bottle Velvet Falernum
1 bottle Luxardo Maraschino
1 bottle passionfruit syrup | 1 litre (1.75 pints) passionfruit purée
2 litres (3.5 pints) mango purée | 2 litres (3.5 pints) lime juice
12 litres (2.5 gallons) apple juice | 12 litres (2.5 gallons) guava juice
60 dashes of Angostura bitters | 25ml (¾ fl oz) absinthe

- Pour 90 per cent of each of the ingredients into a barrel and stir. Taste and balance, then add all the rest anyway. Add a large block of ice and drink. Serves: *everybody*.

20 SEPTEMBER

Orville & Wilbur Wright Fly a Circle in Their Flyer II (1904)

MISCHIEVOUS KEA IPA, ALTITUDE BREWING –

When the Brothers Wright made their first ascension towards the clouds in their shit-scary, ramshackle, wooden-winged machine, could they have predicted how air travel would evolve in the following 100 years? We've gone from the romance and death-defying adventure of their Flyer II to Easyjet's cattle-pen queue at Luton and the wider destruction of the planet.

Far from adventurous, these days air travel is mostly just unpleasant – but, unlike the Wrights, we do at least have the drinks trolley. So, if you have to fly, enjoy a drink. Note, though, that altitude alters flavour, so drinks need to be more intense in order to satisfy our palates. A piece of research from Lufthansa indicated our sensitivity to savoury and sweet drops by around 30 per cent on a plane because the air pressure and a reduction in humidity play havoc with tastebuds. Bitter and sour flavours fare better. This explains why a punchy gin and tonic is a faithful ally in the clouds, and why many airlines are no longer winging it with average choices, instead offering heavily hopped beers and wider whisky and wine selections. Note that no amount of menu improvement on Ryanair can remove the bad taste that experience leaves you with.

In keeping with the theme, we recommend Altitude beers today. Based in Queenstown, New Zealand, the Altitude team brews 'for the adventurous', cares about the planet, and offers up a constantly changing array of craft beers, influenced by the surrounding environment and local ingredients.

---

21 SEPTEMBER

MARY CHUBB IS LEFT DISAPPOINTED (1915)

SALTY KISS, MAGIC ROCK –

Sir Cecil Chubb discovered buying a bunch of massive stones was not the key to marital bliss when his anniversary gift of Stonehenge failed to impress wife Mary. Her response of 'get your rocks off' was far from a come-on, and, needless to say, she offered no salty kiss that night.

Had craft brewers Magic Rock been available back in 1915, they might've compensated with their own Salty Kiss, a traditional German-style *gose*. Flavoured with gooseberry, sea buckthorn and sea salt, it delivers a finish as sour as a disappointed spouse.

Having spent £6,600 on the monument in auction, Chubb eventually relinquished hold over his prehistoric rocks, presenting them instead as a gift to the entire nation.

22 SEPTEMBER

CATHERINE THE GREAT CROWNED RUSSIAN EMPRESS (1762)

RUSSIAN IMPERIAL STOUT –

During her 35-year reign, Catherine crushed internal uprisings, extended Russia's imperial reach into Poland and the Crimea, befriended French philosophers, composed operas, and championed the arts. But the main reason Catherine the Great was great was because she drank stout. Lots of it – and strong, strapping imperial stouts, too, that would blow the faux foam 'Oirish' hats off the heads of Guinness drinkers everywhere.

These were complex, cockle-warming bruiser beers that were smoky and silky, and her penchant for the style developed in the 1700s during a trip to London, where porters and stouts were ubiquitous. Thrale's Entire, an imperial stout from Southwark, South London, was her favourite, brewed strong with lots of preserving hops to withstand the hazardous voyage from London to Russia.

Some of today's brewers still produce imperial stouts that would have satisfied Catherine. Courage Brewery have breathed life back into the classic Thrale's recipe, while Harvey's, based near Brighton in East Sussex, matures its oily, unctuous, ink-black imperial stout for more than nine months.

23 SEPTEMBER

THE SHAWSHANK REDEMPTION IS RELEASED (1994)

STROH'S BOHEMIAN-STYLE PILSNER –

If you've not seen *The Shawshank Redemption*, you really should. It's a wonderful movie – we give it a rating of four and a half lipsticks out of five. People often bang on about the ending, but, for us, the best moment is when Andy scores his pals some ice-cold beer in return for doing a prison guard's tax return. The scene even comes complete with a Morgan Freeman voiceover.

The inmates drink Pabst Brewing's Stroh's Bohemian-Style Pilsner, created by Bernhard Stroh, who came to the US from Germany in 1848, and now brewed by Pabst according to the original recipe.

24 SEPTEMBER

KENTUCKY FRIED CHICKEN OPENS FIRST FRANCHISE IN UTAH (1952)

MEATLIQUOR RADLERITA –

Some enjoy a bit of finger licking after dipping into one of those big ol' greasy buckets at KFC, but top of our fried chicken pecking order is MEATliquor. Their wings are clucking delicious and are a perfect hors d'oeuvre before a Clustercluck burger. As well as fine purveyors of cooked meat products, the restaurants also present excellent beverages and the Radlerita is an ideal quencher.

~

Ice | 50ml (1½ fl oz) tequila blanco
25ml (¾ fl oz) agave syrup | Squeeze of lemon or lime

Stiegl Radler Grapefruit beer
Pink grapefruit slice, to garnish

- Fill a well-chilled 500ml tankard with ice. Add the tequila, agave syrup and lemon or lime juice, then top with the beer and stir. Garnish with a slice of pink grapefruit.

---

25 SEPTEMBER

World Dream Day

SANDEMAN PORT –

If you finish a night off with a bit of port and cheese, understand it's not the Stilton – or indeed the Sandeman – that's bringing you a dream.

Dreams occur during the REM (Rapid Eye Movement) stages of sleep, around two hours after you pass out. If you have a healthy sleep cycle, you transition from REM into lighter sleep as the night continues, meaning when you wake, you don't tend to remember the details of your dreams.

While alcohol gets you into that vivid dream state quicker, as it wears off, your neurotransmitters are agitated and you wake suddenly, invariably interrupting a vision of you, say, sleeping with a neighbour, that sort of thing. So, you lose that transition phase, the critical spell of sleep in which you would otherwise forget the ins and outs, so to speak, of a dream.

If you wake up in the middle of something like that, you tend to remember it, so while it's common to blame drink or cheese for the weaving of a strange sleepy-time, in reality (or non-reality), you spend most nights dreaming about your sexy neighbour – you just have a healthy stretch of sleep after REM to help you bury it all away.

Indigestion creates the same disruption, and since cheese appears at the end of your meal, and is invariably not required, this will double down on disruptive duvet time.

26 SEPTEMBER

DEATH OF PAUL NEWMAN (2008)

FLIP COCKTAIL -

'Nobody can eat 50 eggs.' So said Oscar-winner George Kennedy to Paul Newman's protagonist Cool Hand Luke in the film of the same name. Fifty certainly seems like a lot. Sure, we've eaten 50 eggs in our lives, but in an hour? As the French would say, '*One egg is an oeuf, Luke.*' And, as our mothers might add, 'You don't want to get egg bound.' But Luke ate every last one, and Newman got an Oscar nomination for his performance.

So, while his legacy includes Academy Award nominations in five different decades (including a win for *The Color of Money*), extraordinary philanthropy, a Jean Hersholt Humanitarian Award for charitable work and a range of his own condiments that still donates 100 per cent of its profits to charity, we'll give the egg scene in *Cool Hand Luke* the hard shell, because it leads nicely into this eggy drink.

A flip is a style of drink, and is nicely interchangeable with your chosen spirit if you follow the below method.

~

60ml (2 fl oz) brandy, whisk(e)y, gin, rum, etc.
15ml (¾ fl oz) sugar syrup | 1 egg | Ice

- Shake all the ingredients (except the ice) in a cocktail shaker to get the egg fluffy, then add the ice and shake again. Strain into a Martini glass or coupe.

27 SEPTEMBER

LOCOMOTION NO. 1 BECOMES FIRST PASSENGER TRAIN (1825)

LAST TRAIN OATMEAL STOUT, FOURPURE BREWING CO. –

In 1825, George Stephenson's *Locomotion* No. 1 took the first passengers on a public rail line, the Stockton and Darlington Railway in the northeast of England, travelling from Shildon to Stockton. You can still get from Shildon to Stockton these days, but you have to change in Thornaby, so you're looking at around one hour ten for the journey. An off-peak single will cost you £6.40.

28 SEPTEMBER

PENICILLIN DISCOVERED (1928)

PENICILLIN –

It seems unlikely the fungi we discover about our person will earn us a Nobel Prize. Not so for white-coated boffin and Nobel Prize-winner Alexander Fleming, who accidentally discovered penicillin while he was cleaning up his mucky lab. Proof you should never throw things away.

Bartender Sam Ross's cocktail of the same name isn't quite as revolutionary as the Scot's antibiotic, but this beautiful peaty Scotch sip is rightly hailed as a 'modern classic'. We've adapted it slightly.

~

60ml (2 fl oz) blended Scotch | 20ml (¾ fl oz) lemon juice
20ml (¾ fl oz) honey and ginger syrup (see opposite)
Ice | 5ml (¼ fl oz) Islay single malt Scotch whisky
(double this if you love Islay malts)

- Shake the blended Scotch, lemon juice and syrup in a cocktail shaker with ice, then strain into an ice-filled rocks glass. Top with the Islay whisky and serve.

To make the honey and ginger syrup: Peel and thinly slice 100g (3½ oz) fresh root ginger. Place in a saucepan with 240ml (8 fl oz) runny honey and 240ml (8 fl oz) water over a medium heat. Bring to the boil, then reduce the heat to low and simmer for 5 minutes. Allow to cool, then strain into a bottle. Store in the fridge for up to two weeks.

---

29 SEPTEMBER

JOHN D. ROCKEFELLER BECOMES THE WORLD'S FIRST BILLIONAIRE (1916)

BILLIONAIRE –

Having founded the Standard Oil Company, American oil tycoon Rockefeller was declared the world's first billionaire in 1916. He would later turn to charity and give away $500 million, which is nice, but he didn't drink, which isn't exactly in keeping with this book. Regardless, this drink was created at Employees Only, a cocktail bar in New York.

~

60ml (2 fl oz) bourbon | 30ml (1 fl oz) lemon juice
15ml (½ fl oz) grenadine | 15ml (½ fl oz) sugar syrup
7.5ml (¼ fl oz) absinthe | Ice | Lemon slice, to garnish

- Shake all the ingredients (except the garnish) in a cocktail shaker with ice, then strain into a chilled cocktail glass. Garnish with a lemon slice.

30 SEPTEMBER

CHEERS AIRS FOR THE FIRST TIME (1982)

SAMUEL ADAMS BOSTON LAGER –

Our favourite character from *Cheers*, which ran from 1892 to 1993, was Norm Peterson, an elbow-bending antidote to the excess of 1980s America.

A black-belt in deadpan delivery, Norm had a deep love of American lager – which along with hapless bartender Woody, fuelled some of the finest lines in comedy sitcom history.

Woody: Can I pour you a draft, Mr. Peterson?

Norm: A little early, isn't it, Woody?

Woody: For a beer?

Norm: No, for stupid questions.

Cheers, Season 6 episode 18, written by David Lloyd.
Created by Glen Charles, Les Charles and James Burrows.

OCTOBER

1 OCTOBER

ALEXANDER THE GREAT DEFEATS DARIUS III OF PERSIA IN THE BATTLE OF GAUGAMELA (331 BC)

RUM ALEXANDER –

Born in the 4th century BCE into the barbaric bacchanalia of Macedonia, the world's hardest-drinking nation, Alexander was a great drinker – but he was an awful drunk. After a goblet too many in 328 BCE, Alexander pierced the heart of his childhood friend, Cleitus the Black, with a javelin.

Talking of javelins, Alexander also organised a drinking Olympics in India. But the Indians weren't particularly good at drinking. All the 'athletes' died, making for a disastrous Olympic legacy.

Despite his boozing, Alexander still created the largest empire the world has ever seen, spanning from Greece to India. Not only that, the pocket-sized, epileptic bisexual also introduced a whole load of exotic things to Europe, including cotton, crucifixion, bananas, ring-necked parakeets and, crucially, rum.

After sampling the sugar-based alcohol in India, Alexander exported sugar cane to Europe, calling it 'the grass that gives honey without bees'. So, celebrate his epic Persian victory with this sweet twist on the Brandy Alexander.

~

50ml (1½ fl oz) Diplomatico Mantuano Rum
20ml (¾ fl oz) chocolate liqueur | 12.5ml (½ fl oz) Irish cream liqueur
25ml (¾ fl oz) single cream | Ice

- Shake all the ingredients in a cocktail shaker with ice, then strain into a coupe.

2 OCTOBER

THE SHINING RELEASED (1980)

JACK DANIEL'S –

Colorado's Overlook Hotel, the fictional setting for Stanley Kubrick's *The Shining*, would get a terrible Tripadvisor review.

Not only has it previously hosted mob-style executions and numerous murders, its former caretaker, Delbert Grady, murdered his little girls with a hatchet and shot both himself and his wife. It has bobbly bathroom towels, too.

When Jack Torrance, played by Jack Nicholson, takes over as caretaker, things don't improve. An unfulfilled writer and recovering alcoholic with anger management issues, Torrance is holed up with a seriously pissed-off wife and a tormented supernaturally skilled son who gets chased down a hallway by a fire hose.

Jack, understandably, needs a drink. So he hits the hotel bar, where Lloyd the bartender pours him a shot of Jack Daniel's on ice. But neither Lloyd nor the bar is real, merely a figment of Jack's imagination – an imagination living in a brain suffering from serious mental flip-flaps.

After a few drinks and a quick chat with Grady's ghost, Jack chases his family around the hotel with an axe.

They should have gone to Center Parcs.

3 OCTOBER

GERMANY UNITY DAY

A GERMAN BEER –

After more than forty years of harrowing division, East and West Germany became one in 1990, leading to 'high *fünfs*' all round.

The annual celebrations focus on different cities every time, and, as German beer drinkers have remained steadfastly faithful to their local regional styles, you can switch your beer style every year.

You've got helles and *weissbier* in Munich, *kölsch* in Cologne (see 14 August) and its fierce rival Altbier in Düsseldorf. Then there's Dortmunder, Dortmund's blue-collar beer, while Berlin has sour Berliner Weisse (see 26 June).

Wurst kaas scenario, just have a big sausage and some cheese.

---

4 OCTOBER

NATIONAL VODKA DAY (AMERICA)

CHILLED SHOT OF BELVEDERE VODKA –

'The three most astonishing things in the past half-century were the Blues, Cubism and Polish vodka.' – *Pablo Picasso*

---

5 OCTOBER

BREAKFAST AT TIFFANY'S RELEASED (1961)

BREAKFAST MARTINI –

The scene of Audrey Hepburn's Holly Golightly clutching a coffee in one hand and a Danish in the other, staring serenely at a window full of jewellery, transcends cinema.

Her favourite drink is 'one-half vodka, one-half gin, no vermouth', but we suggest you go for the Breakfast Martini instead, created by Salvatore Calabrese at London's Lanesborough Hotel.

~

45ml (1½ fl oz) gin | 1 tablespoon orange marmalade
10ml (½ fl oz) triple sec | 10ml (½ fl oz) lemon juice | Ice

- Mix together the gin and marmalade in a cocktail shaker until combined. Add the triple sec and lemon juice, then shake with ice and fine-strain into a Martini glass.

---

6 OCTOBER

MAD HATTER DAY

CRAZY GIN & TONIC IN A TEACUP –

Back in the mid-1980s, a bunch of computer geeks from Boulder, Colorado, decided to dedicate a day to silliness by wearing funny hats on 6 October. Crazy guys.

Their choice of date refers to the '10/6' ticket adorning the distinct millinery worn by the Mad Hatter in *Alice in Wonderland*, but the term 'mad as a Hatter' derives from 18th-century milliners who, while making fur hats, were exposed to excess amounts of mercury. This induced Danbury Shakes syndrome, with symptoms including slurred speech, tremors, hallucinations and memory loss. And memory loss.

Anyway, if you really must, put on a wacky top hat today and make yourself a gin and tonic with Crazy Gin, distilled from Indian Lassi. And, of course, serve it in a tea cup.

---

7 OCTOBER

BUDWEISER BUDVAR FIRST BREWED (1895)

BUDWEISER BUDVAR –

The town of České Budějovice is the birthplace of Budweiser.

In the 15th century, it was home to 44 breweries and the royal court brewery of Bohemia.

Its beers were known as Budweisers, and, due to the royal connection, the 'Beer of Kings'. Alas, České Budějovice's brewers neglected to trademark the name, and in 1845, American brewer Anheuser-Busch chose 'Budweiser' as the name of its new yellow lager, even bastardising its regal refrain.

In 1895, when the Budejovice Brewery was founded and began exporting its Budweiser Budvar beer, this miffed those in Missouri. After more than a hundred years, and numerous lawsuits, the two breweries are still arguing over the Budweiser brand name.

Clever-clogs lawyers may argue that Anheuser-Busch's beer was trademarked first, but no courtroom could be convinced it's got more flavour. Loyal to brewing traditions, Budvar ferments in open vessels and cold conditions for three months – a lot longer than its American counterpart.

---

8 OCTOBER

FIRST US PATENT FOR THE MICROWAVE OVEN (1945)

A COLD CAN OF LAGER WITH SOME POPCORN –

The microwave was accidentally created by Percy Spencer, a white-coated boffin studying magnetrons 'n' that.

While standing near a radar set, a chocolate bar in Spencer's pocket melted, leading him to wonder whether that meant he could create a device crap pubs could use to heat up food in the future.

Twenty years later, in 1945, an American company more used to making missiles, patented Spencer's microwave cooking process, which was first trialled using popcorn. Spencer then tried microwaving an egg, but it exploded on a poor technician's face, leaving him unharmed – if a little shell-shocked.

9 OCTOBER

BIRTH OF JACQUES TATI (1908)

BRETON CIDER –

Do you need cheering up? Do you hanker after simpler, more straightforward, sepia-tinted times? Then watch *Les Vacances de Monsieur Hulot*, a 1953 classic French comedy mocking the dogged determination of people trying to enjoy themselves on holiday.

Set in Brittany, home to some lovely cider, its protagonist is a clumsy buffoon who unsettles an entire resort with his accident-prone antics while dressed in trademark droopy hat, raincoat, pipe, ankle-swinging slacks and umbrella.

Jacques Tati, who wrote the film and played the lead role, was skilled in the art of slapstick and cutting satire that mocked the perils of modernity and the drudgery of everyday life.

'Everybody is sad,' Tati once said, probably while puffing a Gauloises. 'Nobody whistles in the streets anymore. I want to proclaim the survival of the individual in a universe that is more and more dehumanised.'

10 OCTOBER

DR. NO RELEASED (1962)

VODKA MARTINI –

Dr. No contains the first scene featuring James Bond drinking his signature vodka Martini. He has two Martinis during the film, both medium dry, both shaken not stirred, one with a lime garnish and the other with lemon – and both made using Smirnoff (the first of many Bond product placements).

The phrase 'shaken not stirred' features approximately 20 times throughout all the films released to date, and while most bartenders scoff at Bond's preference, shaking rather than stirring suited the secret agent, giving him more time to muck about on jetpacks and wheedle his way into women's knickers.

Not only particular about how his Martini was made, Bond was also super specific about the V-shaped vessels in which it was served. In fact, so obsessed was he with glasses, he opened an optician business in south London after retiring from the Secret Service. It was called 'For Four Eyes Only'.

~

60ml (2 fl oz) Smirnoff vodka | 15ml (½ fl oz) dry vermouth | Ice
Lemon zest twist or olive, to garnish

- Stir the vodka and vermouth in a mixing glass with ice, then strain into a Martini glass. Garnish with an olive or a lemon twist.

---

11 OCTOBER

MEDITRINALIA

BAROLO WINE –

Today, we're reviving Meditrinalia, a long-forgotten Roman festival celebrating the end of the grape harvest. This involves drinking new wine mixed with old in honour of the god Jupiter. Jupiter's son was Bacchus, the chubby, mischief-making god of wine, who gave us the term 'Bacchanalia'.

Now, Bacchus' birth was a bit unusual. He was initially conceived when Jupiter had sexy time with Semele, a human. So, a god and a human had a child, which is unusual in itself. But things got weirder when Jupiter accidentally exposed himself as a divine being – and she spontaneously combusted. We've all done it.

Jupiter rescued the embryo (Bacchus) and, doing what any dad would do, sewed him onto his thigh. Nine months later, Bacchus was born – an event represented by the annual pruning of the vines.

What a carry on.

12 OCTOBER

National Day of Spain

OLOROSO SHERRY –

To celebrate the National Day of Spain, we've written a cliche-ridden poem for you. You're welcome.

San Miguel and cheap Cruz Campo, Cava wine and vino tinto
And don't forget the jugs of San-gri-ah
From Barcelona to Cordoba, drink rioja around the clock-a
And pair it all with lovely paella
As you click your castanets, stuff your face with big croquettes
And drink dry sherry from a wooden bodega
Salute the great El Cid while watching Real Madrid
And finish with a six-hour siesta.

13 OCTOBER

Lenny Bruce Born (1925)

C*CK-SUCKING COWBOY –

Lenny Bruce was a wry revolutionary whose free-form, fast-flowing 60s comedy railed against the Vietnam War, racism, censorship, organised religion and any kind of bigotry that got his back up.

'I'm not a comedian, and I'm not sick,' he once said, in his thick New York Jewish drawl. 'The world is sick, and I'm the doctor. I'm a surgeon with a scalpel for false values.'

Bruce introduced scathing social commentary into the art of stand-up, but his routines regularly featured very rude words indeed and he was banged up several times on obscenity charges.

The first time was in San Francisco, in 1961 (for saying 'c**k-sucker'), but the routine that ultimately led to a high-profile obscenity trial and a four-month jail sentence featured gags about sex with a chicken.

Forbidden from performing, it all became too much for an increasingly dispirited Bruce and he overdosed on morphine aged just 40. He remains the only American comedian to be jailed for his words, but by the time he was pardoned in 2003, he had influenced a whole generation of comedians, including Richard Pryor, Bill Hicks and Jerry Seinfeld. Raise a glass to him today.

~

15ml (½ fl oz) butterscotch schnapps
30ml (2 fl oz) Baileys Irish Cream

- Layer the two ingredients in a shot glass and enjoy.

---

14 OCTOBER

PREMIER OF KUNG FU TV SERIES (1972)

GRASSHOPPER –

Do you remember being karate-chopped in the playground by one of the bigger boys as he recited Taoist maxims in a racially-insensitive Chinese accent? We do.

It was all down to *Kung Fu*, a classic martial arts action drama that espoused ancient Taoist philosophy and inspired a boom in oriental martial arts. It starred David Carradine as Kwai Chang Caine, a Shaolin master on an odyssey through the American West in search of his long-lost half-brother.

He is schooled by an old, blind master called Master Po, who refers to his apprentice as 'Grasshopper'. So have a classic minty cocktail of the same name, which will fulfil the age-old Taoist truism: 'Stop thinking and end your problems.'

~

25ml (¾ fl oz) crème de menthe | 25ml (¾ fl oz) crème de cacao
25ml (¾ fl oz) single cream | Ice

- Shake in a cocktail shaker with ice, then strain into a coupe.

15 OCTOBER

Breast Health Day

OCHO REPOSADO TEQUILA –

Breast Cancer is the most common cause of cancer in women worldwide, so it's vital you regularly check your 'shirt potatoes'.

One woman well-versed in mindful mammary maintenance is Mayahuel, the Aztec goddess of Agave. Mayahuel had 400 breasts that fed 400 children with *pulque*, the fermented alcoholic sap of the agave plant, which is used to make tequila.

Following an ill-advised amorous adventure with the god of medicine, Patecatl, Mayahuel's offspring, for some reason, turned into rabbits – each one representing a different form of intoxication – and this is why people say tequila puts hares on your chest.

Anyway, tequila may have its knockers, but we love it – especially 100 per cent agave tequilas like Ocho Reposado Tequila, an acutely traditional tequila aged in ex-American whiskey barrels for 8 weeks and 8 days.

16 OCTOBER

Birth of Oscar Wilde (1854)

GREEN FAIRY –

Raise a glass to one of literature's wittiest figures filled with some Wilde 'Green Fairy' liquid.

Absinthe shared its hue with the green carnation, famously worn on Wilde's lapel as a well-known symbol of homosexuality. 'Absinthe has a wonderful colour, green,' wrote Wilde. 'A glass of absinthe is as poetical as anything in the world. What difference is there between a glass of absinthe and a sunset?'

~

35ml (1 fl oz) absinthe | 35ml (1 fl oz) lemon juice
35ml (1 fl oz) water | 25ml (¾ fl oz) sugar syrup
Dash of Angostura bitters | Ice

- Shake all the ingredients in a cocktail shaker with ice. Strain into a chilled coupe.

---

17 OCTOBER

Birth of Evel Knievel (1938)

A SHOT OF WILD TURKEY BOURBON –

Evel Knievel famously broke every bone in his body in a daredevil career consisting of more than 300 death-defying jumps – and three years in hospital.

Knievel successfully landed hundreds of jumps, but most of his famous feats ended in crashes. In 1968, wearing his trademark red, white and blue jumpsuit, he jumped over the fountains at Caesars Palace Hotel, Las Vegas, misjudged his landing, breaking his skull, hip, pelvis and ribs. He was in a coma for a month.

When he woke up, he carried on jumping, crashing and becoming richer and more famous. In 1974, Americans turned on their TVs to see if he could jump Idaho's Snake River Canyon on a rocket-powered motorcycle. He couldn't. He had to jump out in a parachute and broke his nose.

A year later, 90,000 people packed into Wembley Stadium to witness Knievel jump over 13 buses for $1 million. He crashed and broke his pelvis. 'Anybody can jump a motorcycle,' he said. 'The trouble begins when you try to land it.'

Prior to rebelling against the laws of gravity, Knievel would often take a small sip of Wild Turkey, which he kept in a hollowed-out cane. He famously mixed it with beer and tomato juice to create his favourite drink, a Montana Mary – which sounds more dangerous than the jumps themselves.

Knievel died at the age of 69, and at his funeral, Hollywood actor

Matthew McConaughey delivered the eulogy. 'He's forever in flight now. He doesn't have to come back down. He doesn't have to land.'

Which is probably no bad thing.

18 OCTOBER

National Cravat Day

GARDEN BREWERY PILSNER –

The modern man's burden is real. Every morning, before embarking on yet another tedious day of deferential drudgery wriggling under the thumb of 'The Man', we place our heads inside a cloth noose (also known as a tie).

It's the fault of 17th-century Croatian soldiers who, while fighting for Louis XIV's French army, wore a slackly tied, lace-edged linen cravat. The French, who always love a dandy, even in battle, christened it a Cravat, a corruption of the Gallic word for Croat.

Why do we wear ties? Well, anthropologists reckon its arrow-like shape is designed to surreptitiously direct one's attention downwards, towards the tie-wearer's genitals. Others reckon it's to cover-up unsightly buttons.

Either way, charm your neck with this pilsner from a cool craft brewery in Zagreb.

19 OCTOBER

US Embargo on Cuba Exports (1960)

BACARDÍ AND COKE –

Before Facundo Bacardí began distilling in Santiago, Cuba, back in 1862, rum was rough gear synonymous with salty sea dogs.

But Bacardí, a Spanish immigrant, succeeded in sophisticating the spirit, making it lighter, easier to drink and classically Cuban –

fuelling Havana's reputation as the party capital of the Caribbean. During Prohibition, thirsty Americans drank Cuba dry.

Passionate proponents of Cuban independence, the Bacardí family backed Fidel Castro's fight against the US-backed dictator Fulgencio Batista in the 1950s, even buying and supplying weapons.

But after Castro took power, he seized all their assets and kicked the Bacardí family out of Cuba. Fortunately, having been distrustful of the Batista regime, Bacardí had set up an array of independent businesses outside of Cuba.

This allowed Bacardí to become a household name all over the world – apart from, of course, Cuba, where Havana Club claims to be the true Cuban rum. It was first produced by the Arechabala family in the 1930s, but after they too were exiled, Havana Club was nationalised in 1959. A fierce rivalry between the two rums has rumbled on for decades, becoming even more heated in the mid-1990s when the Cuban government joined forces with drinks giant Pernod Ricard and made their Havana Club a global player.

A year later, the Arechabala family passed its Havana Club recipe and production techniques to the Bacardí family – who now make their own Havana Club in Puerto Rico.

Since Castro's death and the thawing of the 1960 American-Cuban embargo, Bacardí's potential return to Cuba has been mooted amid increasingly intensive legal battles between the two rums.

We've been strongly advised by our legal team to not get involved.

20 OCTOBER

SYDNEY OPERA HOUSE OPENED BY QUEEN ELIZABETH II (1973)

PACIFIC ALE, 4 PINES BREWING COMPANY (SYDNEY) –

Designed by Danish architect Jørn Utzon, the Sydney Opera House is Australia's most famous landmark.

It regulates its temperature using seawater drawn directly from the harbour below. When the Sydney Symphony Orchestra is in residence, the temperature must be 22.5°C to prevent its musical instruments being affected by humidity. We could mention moist G-strings here … but we're better than that.

21 OCTOBER

DEATH OF ADMIRAL LORD HORATIO NELSON (1805)

NELSON'S BLOOD NO.2 –

Not content with having a huge column, which birds still flock around today, Admiral Lord Horatio Nelson is famous for giving Napoleon's French Navy a proper spanking at the Battle of Trafalgar.

A highly skilled commander schooled at naval college (where he also learned to belly-dance), Lord Nelson used cunning tactics to beat the French (22 ships to nil), but he was sadly killed by a French musketeer.

His body was preserved in a barrel of rum that was drunk dry by his adoring crew on the voyage home – giving rise to 'Nelson's Blood', the naval term for rum. Pusser's Rum is a faithful replica of the naval rum ration, last issued on Black Tot Day, 1970 (see 31 July).

~

45ml (1½ fl oz) Pusser's Navy Rum
45ml (1½ fl oz) cranberry juice
20ml (¾ fl oz) lime juice | 20ml (¾ fl oz) orange juice
10ml (½ fl oz) sugar syrup | 2 dashes of Angostura bitters
Ice | Lime wedge, to garnish

- Shake all the ingredients (except the lime wedge) in a cocktail shaker with ice, then strain into a tall ice-filled glass. Garnish with a lime wedge to prevent scurvy.

---

22 OCTOBER

THOMAS EDISON INVENTS A COMMERCIALLY PRACTICAL, LONG-LASTING LIGHT BULB (1879)

CALIMOCHO (EQUAL PARTS COLA AND CHEAP RED WINE) –

Thomas Edison drank Vin Mariani, a 'medicinal' mix of red wine and coca leaves (see 17 December). Believed to be the precursor to Coca-Cola, it contained 21mg cocaine per 100ml (3½ fl oz). Edison drank it to stay awake at night – which is why he needed the light bulb.

Edison also believed that there were a dozen or so miniscule people living in each of our brains. When we die, he claimed, these little people simply pack their bags and go and live in someone else's head.

This is what happens if you do too much cocaine …

23 OCTOBER

NATIONAL HUNGARY DAY

TOKAJI WINE –

Hungary has given us the Rubik's cube, the biro, Tony Curtis, Harry Houdini and the only political constitution drawn up on an iPad.

It's also home to the world's first official wine region, Tokaj, which has been producing Tokaji wine since the 5th century. It offers a slippery sweet sip that, when poured into a glass, laces more than a shoe shop-assistant on speed.

24 OCTOBER

WOMAN GETS SHOT BY A DOG CALLED TRIGGER (2015)

TRIGGER PALE ALE, MUSKET BREWERY –

Allie Carter was hunting waterfowl in Indiana when she momentarily placed her shotgun on the floor. Her chocolate Labrador stepped on it, and inadvertently shot her in the foot. To add insult to minor injury, the dog's name was Trigger.

25 OCTOBER

BATTLE OF AGINCOURT (1415)

LA BAVAISIENNE AMBRÉE, BRASSERIE THEILLIER –

The Battle of Agincourt was a clash between two kings in crisis. The tenure of England's Henry V was being threatened by a restless nobility, while in France the aristocracy were also looking to overthrow Charles V, who was clearly a couple of Camemberts short of a *fromagerie*.

Realising nothing united England quite like administering the French a proper kicking, Henry went to France for some fisticuffs. He won the first battle in Harfleur, but his 12,000 strong army was decimated by injury, deaths, desertions and rampant dysentery.

Sick, famished, frail and outnumbered by as many as six to one, the English were met in Agincourt by a fully rested French force vowing to cut off the fingers of every English archer they captured.

After a cagey opening, Henry surprised his opponents by moving his troops up the quagmire pitch in Picardy. When the flustered French responded with a cavalry-led counter-attack, England's 6,000 rapid-firing archers peeled back their longbows and picked off their opponents with ease. The sky rained with arrows, mallets, axes and swords, massacring 10,000 French soldiers in the mud.

In a wholly unexpected away win for the English, they only lost 400 soldiers, and Henry returned to London a national hero, his claim to the throne legitimised.

After securing Normandy, however, Henry's ambitions to take the French throne were scuppered by a severe case of the squits, which killed him in 1422.

Honour Henry with this brilliant 'Bière de Garde', an acutely idiosyncratic saison-style beer rooted in the rustic regions of northern France, an arrow's flight from Agincourt.

---

26 OCTOBER

GUNFIGHT AT THE O.K. CORRAL (1881)

OLD OVERHOLT AMERICAN WHISKEY –

The Wild West's most famous, rootin'-tootin' gunfight took place between two rival factions in the frontier town of Tombstone.

On the law-enforcement side, you had Wyatt Earp, his brothers

Morgan and Virgil, and their friend Doc Holliday, a gambling, gunslinging dentist. On the other? Two pairs of outlaw, son-of-a-gun, cowboy siblings: the McLaury brothers, and Ike and Billy Clanton.

Their beef began, quite literally, when the Earps accused the McLaurys of stealing horses and donkeys and selling them as 'cows' to the local Tombstone butchers (whose relaxed meat standards simply wouldn't be allowed today).

Steaks were raised further when Holliday and Billy Clanton came to blows in the local saloon. After the scuffle, Clanton vowed he'd hand Holliday his ass on a plate (which is exactly what the local butcher had been doing).

Anyhow, the next morning, after failed attempts by the local sheriff to calm the situation, the two factions came face-to-face in a lot near the Old Kindersley Corral.

No-one knows who fired the first shot, but after a flurry of gunfire, as the smoke cleared, three men (the McLaury brothers and Billy Clanton) lay dead. It only lasted half a minute, but the gunfight became a fateful part of frontier folklore. As we like to remind our wives, some of life's most memorable moments needn't last more than 30 seconds.

Pay homage with the favoured whiskey of Doc Holliday, whose famous dying words, as he sipped his glass of Old Overholt, were, 'This is funny.'

(He wasn't reading this.)

---

27 OCTOBER

DEATH OF LISE MEITNER (1968)

QUANTUM STATE SESSION IPA, ATOM BEERS –

Lise Meitner discovered nuclear fission, which led directly to the advent of nuclear energy.

However, few will have heard of one of the 20th-century's greatest physicists. As a Jewish woman in the 1940s, Meitner has been airbrushed from history, and, like so many other women scientists, her work has been shamefully misattributed to male colleagues.

Born in Vienna in 1878, Meitner studied at Berlin University, where she met fellow physicist Otto Hahn. As women weren't allowed into the university's chemistry institute, she worked unpaid in an underground ex-carpenter's shed and was refused an official academic position.

Meitner spent the First World War on the front lines as an X-ray nurse with the Austrian army before carrying out ground-breaking global nuclear research throughout the 1920s and 1930s.

Having been forced to flee Nazi Germany in 1938, long after her fellow Jewish counterparts Albert Einstein and Erwin Schrödinger had left, Meitner made her epic breakthrough while exiled in Copenhagen in 1939. In accordance with Einstein's $E = mc^2$ equation, she discovered how to split the atom and shared her discovery with Hahn. But when he published his paper, Meitner wasn't mentioned, because Hahn realised any acknowledgement of a Jewish scientist would end his career.

When he won the Nobel Prize in Chemistry in 1944, Hahn ignored Meitner's instrumental role in discovering nuclear fission, and Meitner spent her remaining career working in rudimentary research facilities in Sweden.

Having experienced first-hand the evils of conflict during the First World War, she steadfastly refused to join America's atomic bomb project, appalled that her 'creation' had become a weapon of evil.

By the time she died in 1968, she had been nominated for the Nobel Prize on 48 different occasions – but never received it.

28 OCTOBER

Birth of 'Garrincha' (1933)

CAIPIRINHA –

If Pelé is Brazil's most famous footballer, Garrincha is the most loved. Manuel Francisco dos Santos, born in 1933 with a deformed spine, had a bent left leg two inches shorter than his right, which bent the other way. His wonky gait and slim frame, resembling that of a small bird, earned him his nickname: 'Garrincha', meaning 'the wren'.

One of the most unpredictable and exciting footballers in the world, he was a wizard of the dribble who brutalised full-backs for fun. While wearing Brazil's canary-coloured shirt, he scored 34 goals and won two World Cups, in 1958 and 1962.

Yet his career, like so many flawed footballing geniuses, was undermined by the Brazilian's off-pitch Bacchanalian lifestyle. He allegedly lost his virginity to a goat, aged 14, and regularly got his bent leg over, fathering 14 children and having numerous affairs with glamorous Brazilian showgirls.

He liked a drink, too. If anything, a bit too much. He tragically killed his mother-in-law in a drunken car crash, and regularly enjoyed cachaça, the Brazilian sugar-cane rum, for breakfast.

Have some in a Brazilian Caipirinha, pronounced 'ky-per-rean-yah'.

~

60ml (2 fl oz) cachaça | 15ml (½ fl oz) sugar syrup
1 lime, cut into wedges | Ice

- Muddle the lime in a sturdy rocks glass, extracting the oils and juice. Pour the cachaça into the glass, along with the sugar syrup, then top up with crushed ice and stir.

29 OCTOBER

Republic Day (Turkey)

EFES PILSNER –

There are loads of things to like about Turkey. The baths, the heart-attack coffee, those cool shoes that curl up at the end, and, also, the world's oldest pub.

Göbekli Tepe, an archaeological site regarded as the world's oldest building, was built before the wheel was invented. More than 12,000 years old, it pre-dates Stonehenge and Egypt's pyramids by 6,000 years.

Archaeological experts initially reckoned it was solely built as a temple, but having since discovered some ancient tubs containing traces of fermented grain, they now reckon that this collection of columns and pillars was, in fact, an early example of a pub.

By modern standards, it's a lousy local: no roof, no toilets, no crisps – it doesn't even have a fruit machine. But there is a sign above the bar, saying: 'Göbekli Tepe – we may not have much stuff-in, but you'll still find us in Turkey.'

30 OCTOBER

Rumble in the Jungle (1974)

THE UPPERCUT –

Muhammad Ali arrived in Zaire to fight George Foreman with his boxing career hanging in the balance.

Indestructible during the 1960s, Ali was in his thirties and ring-rusty following a lengthy ban for refusing to fight in Vietnam. In 1971, he'd suffered his first professional loss to Joe Frazier, and Ken Norton broke his jaw two years later.

The undefeated Foreman, meanwhile, had won all but three of his forty fights within three rounds, and made light work of both Norton and Frazier. No one gave Ali a chance of loosening Foreman's grip on the heavyweight division.

But, as ever, Ali got into Foreman's head before they even got in the ring, mocking his plodding style and calling him 'The Mummy'. Hailed a hero of Black rights by the people of Zaire, Ali also deliberately turned his African hosts against his opponent, whipping up crowds and getting them to chant 'Ali, *bomaye*' ('Ali, kill him') wherever he went.

By the time the first-round bell was sounded, Foreman was absolutely furious, and he got even angrier when Ali landed a series of outrageous right-handers on his head.

Foreman sprung off his stool for the second round, seething with anger, only to find Ali retreating to the ropes, inviting Foreman to come to him. It seemed a suicidal tactic given Foreman's punching power, but Ali, covered up and clinging on to Foreman, survived the round, taunting him throughout.

For the next four rounds, an increasingly enraged Foreman furiously went to work on Ali's head and torso but, having never fought beyond the third round, he began to physically tire, the weight of his punches waning with every jab.

In the seventh round, with his 'rope-a-dope' strategy working and Foreman all punched out, Ali delivered a devastating psychological blow. When Foreman unleashed a heavy upper-cut to Ali's jaw, Ali gently whispered in his ear: 'That all you got, George?'

In the next round, Ali felled a broken Foreman with a plum right-hander to take back the heavyweight title of the world, completing one of the most remarkable comebacks in sporting history.

~

50ml (1½ fl oz) Chivas Regal 12
25ml (¾ fl oz) pineapple juice | 25ml (¾ fl oz) lemon juice | Ice
Fresh mint sprig and pineapple slice, to garnish

- Build the whisky and juices over ice cubes in an Old Fashioned glass and stir. Garnish with a mint sprig and pineapple slice.

31 OCTOBER

HALLOWEEN

PENDLE WITCHES BREW, MOORHOUSES BREWERY –

Rather than spending the evening on your doorstep, handing out sweets to badly dressed, troublesome youths, we suggest you usher in the gloomy, cold days of winter with this lovely Lancastrian bonfire-hued ale, named after some witches convicted and hanged back in 1612.

NOVEMBER

1 NOVEMBER

Seabiscuit Wins the 'Match of the Century' (1938)

SIETE LEGUAS TEQUILA –

In a true under-horse story, Seabiscuit, the diminutive and dodgy-kneed nag, took on War Admiral, the strong and swift stallion everyone expected to storm to victory. But against all odds, Seabiscuit won, the nation went nuts and bookies' pencils fell limp.

Seabiscuit hasn't lent his name to a drink – it fits better with custard creams created by the coast – but another horse, Siete Leguas, has, and that'll do for us.

Translating to Seven Leagues, this mighty Mexican steed carried Pancho Villa into conquests, with the revolutionary charging his charger with a pre-battle shot of tequila.

Today, Siete Leguas tequila is produced using a horse-drawn *tahona*, a huge volcanic stone wheel that rolls over and crushes baked agave before the juice is fermented and distilled. The blanco is 100 per cent agave spirit. It tastes fresh, with a sweet agave note that makes it great neat or in a quality cocktail like a Tommy's Margarita.

2 NOVEMBER

Birth of Marie Antoinette (1755)

CHAMPAGNE –

While Marie Antoinette probably didn't ask peasants to ram cake down their gobs, it's equally unlikely the champagne coupe glass is based on the contours of her breasts.

While the idea will provoke a little titter as you pour a champagne into a coupe, weigh up the complications: on the one cupped hand, who would let a lowly glass blower near her bits? And on the other, what about the stems?

More exposing for this massive mammary mistruth is the fact champagne coupes appeared fifty years before her reign.

But, hold up, wait a minute. While the design idea has its knockers, it is not without historical support.

In 1787, Jean-Jacques Lagrenée le Jeune and Louis-Simon Boizot really were given access to Marie's personal space and subsequently designed the *Bol Sein*, or *Jatte tétons*, which translates literally to 'nipple bowl'. Marie used the bowls for dainty drinking at the dairy of Rambouillet. So there is a link.

And there's more. The ancient Greeks drank wine from paraboloid-shaped vessels called *mastos* cups, and mastos is Greek for breast, because they were indeed based on said shape, complete with a nipple at the bottom.

So, while this historic queen's legacy should go beyond glasses shaped like jubblies for her bubbly, this entry is far from booze-related booby bobbins.

---

3 NOVEMBER

LAUNCH OF SPUTNIK 2 (1957)

BOTTOM SNIFFER DOG BEER, WOOF & BREW

Dogs in space? You must be barking.

Seriously, that just happened.

But so did a dog in space, because, with tails wagging after the success of Sputnik, the Russians launched Sputnik 2, placing marvellous mutt Laika at the controls.

The Russians reported that Laika liked it a lot, and spent days orbiting our planet, but this was propaganda. Sadly, she died hours into the flight due to extreme temperatures.

So, if you've got your own courageous canine, put your pet first today with a Woof & Brew Bottom Sniffer Dog Beer, a non-alcoholic and non-carbonated drink. The name refers to a dog's exceptional sense of smell, which is up to 100,000 times more sensitive than that of humans, and bottom-smelling, which is a dog's way of saying hello. Given their ultra-sensitive sense of smell, they might be better off saying hi from a distance.

4 NOVEMBER

BARACK OBAMA VOTED INTO THE WHITE HOUSE (2008)

312 URBAN WHEAT ALE, GOOSE ISLAND –

Hailing from Chicago, Obama has described Goose Island as 'a very superior beer,' and mentioned an affection for their 312 Urban Wheat Ale. This hazy, unfiltered beer is another successful American interpretation of the wheat style, with hints of a pale ale thanks to a spicy whack of USA Cascade hops.

5 NOVEMBER

BONFIRE NIGHT

SMALL STEPS DOUBLE DRY HOPPED PALE ALE, THREE HILLS BREWING –

Bell-ends all across the UK will be launching bangers from their bottoms tonight, attempting to emulate the fate of Guy Fawkes, a man burned to death for his bad behaviour.

Putting aside the cruel irony of marking a botched attempt to blow up the Houses of Parliament by blowing things up, it's worth noting the Gunpowder Plot was actually conceived in a pub. Robert Catesby hosted his fellow conspirators in a room above the gate house in the Olde Coach House in Northamptonshire.

Northamptonshire-based Three Hills Brewing serve up this Double Dry Hopped Pale, which works here since it's described as a 'dry-hopped juice bomb'. Plus it comes in cans, so you can sink it down the park while you let off bangers.

---

6 NOVEMBER

IDRIS ELBA VOTED *PEOPLE MAGAZINE'S* SEXIEST MAN ALIVE (2018)

FIVE POINTS PALE ALE,
FIVE POINTS BREWING CO. –

Just as we brushed over the Miss World contest (19 April), we shan't draw attention here to a contest that sexually objectifies men, encourages unrealistic physical appearance goals, muscle dissatisfaction and makes everyone feel guilty about eating crisps with our pint. Sort it out, *People Magazine*. An excellent actor, Idris hails from Hackney, so opt for a beer from his neighbourhood brewery.

7 NOVEMBER

BIRTH OF VLADIMIR SMIRNOV/ SMIRNOFF (1875)

SMIRNOFF –

While sadly no longer with us, Vladimir Smirnov (later Smirnoff) was a man who cheated death on many occasions.

Vlad inherited a booming vodka business from his father Pytor, but World War, Russian prohibition and then Russian Revolution, sapped the Smirnov empire, and by the time the Bolsheviks targeted the capitalist pigs in 1918, Vlad was destitute. To make matters worse, he got himself arrested and sentenced to death.

But rather than the standard one time you get shot at by loads of soldiers, Vlad faced the fusillade an incredible five times – and survived.

He accomplished the first evasion by convincing his executioners to sing with him as they marched towards the firing range. So engrossed in the song were the soldiers, that they merrily skipped beyond the designated area. By the time they realised their error, it was late and they were hungry, so they all went home. The song must've been *really* good.

Vlad then endured a series of mock or postponed executions before being rescued by sympathisers and escaping to Turkey with his wife.

With little to his name except his name, Vlad dropped the 'v' and added a couple of 'f's, and took a stab at selling Smirnoff there. He failed – the Turks have raki, after all. He scraped enough together to reach Bulgaria, where they don't have raki but do have rakia, which is quite similar. So, after failing again, he moved to Poland, where his attempt at selling vodka was a bit like bringing coal to Newcastle – or, indeed, vodka to Poland.

He finally settled in Paris, and while the French like a bit of front, Vlad's insistence his liquid could better *vin rouge* was arrogance too far. So, they said '*non*' and subsequently slammed the open market door in his face.

By the early 1930s Vlad was once again impoverished and close to death when the Ukrainian–American Rudolph Kunnett came to his aid, buying up the rights of his name before taking the vodka to America. It would eventually flourish under the stewardship of American John Martin, president of American alcohol company Heublein, to become one of the biggest brands in drink.

8 NOVEMBER

HITLER LAUNCHES HIS BEER HALL PUTSCH (1923)

LÖWENBRÄU –

Hitler was a teetotaller. Obviously, this wasn't his worst crime, but he was sober when he launched his *putsch* from the Bürgerbräukeller beer hall. That is to say: he came up with that idea sober. So, without being glib about the unspeakable atrocities that ensued, Hitler proves that, if you're in the pub without a pint today, and you have a truly abhorrent idea – like becoming a Nazi – have a drink and have a rethink.

9 NOVEMBER

HARRY HOUDINI ESCAPES (1903)

TETLEY'S NO.3 PALE ALE BEER –

Harry Houdini toured the UK in 1903, and one of his spectacles included escaping from a military prison in which Oliver Cromwell held his prisoners.

However, he wasn't so lucky in 1911. The escape artist came face to face with the fearsome fungus when Joshua Tetley challenged him to tweak his famous milk churn escape trick by using a Tetley's ale barrel instead.

When the milk churn was filled, the sneaky Houdini survived thanks to a secret pocket of air – but in the case of beer, the fermentation process meant yeast expelled CO2 into the gap, denying him oxygen. The air pocket filled with gas, a bit like a Dutch oven, and Houdini started to panic, almost passing out.

Having survived, he might have enjoyed a pint of Tetley's, but Houdini happened to be a teetotaller. Perhaps his lack of brewing knowledge is what led to his error. If he'd drunk Tetley's beer instead of Tetley's tea, he might've nailed the trick.

---

10 NOVEMBER

BIRTH OF WILLIAM HOGARTH (1697)

GIN –

William Hogarth's 1750s *Gin Lane* print really did a number on gin. In the piece, Madame Genever – or Judith Defour – dominates, ginned up to her eyeballs as she lobs her baby into the doorway of a gin shop. Behind her, a coffin is being filled with another woman, whose child watches on, while sipping gin. To the right of the image, a woman adopts a seemingly caring cuddle with her baby – but is force-feeding it gin. There's a dude in the bottom corner who looks emaciated, possibly dead, but has a very fancy gin Martini glass on him, and his dog looks sad because all the gin has run out. There's also a chap who looks like an olden-day hipster. He is drunk on gin and has speared a baby. Buildings crumble, riots explode and starvation ensues. London is gin-soaked and properly done in.

But, while we can all agree children guzzling gin is bad news bears, Hogarth's satirical sketch had a deeper meaning. Gin takes the blame for London's downfall, and certainly it deserves some –

if only Londoners could've drunk less, but better. But Hogarth was identifying a brutal poverty that encouraged the public's escape into alcohol. Aggressive urbanisation and lack of space in London led to the rapid spread of disease, while prostitution, unemployment and neglect of children were commonplace. The pressures on society were greater than the simplicity of a gin epidemic.

---

11 NOVEMBER

PILSNER URQUELL FIRST SERVED (1842)

PILSNER URQUELL –

Imagine it's 1842. You're the Bavarian brewer Josef Groll and you've just invented Pilsner. You're really excited about it, and you take your new invention to a meeting with potential investors.

Up until now, all beer has been dark, dreary-looking and drunk from pewter tankards. But Pilsner, brewed in the town of Pilsen, is the world's first golden beer. It glistens, it glows, it flickers like a flame and dances in clear glassware.

The investors are impressed. They want to drink it; they know everyone else will want to drink it. But just as they are ready to give up the groats, they give up the ghost. Why? Well, you haven't patented the idea, have you? It seems like an obvious bit of admin, but you've dropped the ball. You dufus. Always patent the idea. That's the number one rule, even if the invention is rubbish – like nappies for dogs.

While it seems remarkable these days, the Pilsen-based inventors of Pilsner genuinely didn't trademark their beer. They tried to in 1899, but, by that time, the horse had bolted, and everyone who could make Pilsner was making Pilsner.

More than 100 years on, Pilsner is brewed all over the world – with varying degrees of success and corner-cutting. Pils has, sadly, become a much-abused beer style, a byword for lacklustre lagers and a far cry from the exceptional original: Pilsner Urquell.

12 NOVEMBER

C.H. MIDDLETON PRESENTS *IN YOUR GARDEN* (1936)

SECRET ENGLISH GARDEN COCKTAIL –

In 1936, C. H. Middleton tended a small garden in the grounds of Alexandra Palace for the broadcast of the BBC's first gardening show.

Did you know, a spell of vigorous gardening can burn off 200–500 calories in an hour? So get to it – then you'll deserve this drink.

~

2 lemon slices | 2 apple slices (disc)
50ml (1½ fl oz) Bombay Sapphire English Estate gin
25ml (¾ fl oz) cloudy apple juice | Ice
75ml (2½ fl oz) ginger ale | Lemon thyme sprig

- Place one lemon slice and one apple slice in the base of a highball glass, then pour in the gin and cloudy apple juice and press lightly. Fill the glass with ice, then top with ginger ale and gently stir. Garnish with the other lemon and apple slices, and a sprig of lemon thyme.

13 NOVEMBER

FIRST TINNED PINEAPPLE ARRIVES IN THE UK FROM HAWAII (1895)

ALGONQUIN –

In honour of this momentous day, a pineapple fact for you: Pineapples ripen upside down.

You're welcome.

More pertinent is the fact they are the emblem for the hospitality industry, and you'll occasionally find bartenders wearing a pineapple pin. In the 1920s, Frank Case, the owner of the Algonquin hotel in New York, extended some of that hospitality to Dorothy Parker and the literati members of the Vicious Circle club, providing them with their own table, dedicated waiters and free hors d'oeuvres every lunchtime.

~

50ml (1½ fl oz) rye whiskey | 25ml (¾ fl oz) pineapple juice
25ml (¾ fl oz) vermouth | 2 dashes of orange bitters
Ice | Pineapple slice, to garnish

- Shake all the ingredients (except the garnish) in a cocktail shaker with ice, then strain into a cocktail glass. Garnish with pineapple.

---

14 NOVEMBER

FIRST MARATHON DU MÉDOC (1985)

MÉDOC WINE –

When Paula Radcliffe stopped in the middle of a marathon to do a poo it was pretty funny, but also necessary, because marathons can take a terrible toll on the body.

Runny nose, cramp and blisters are all obvious pitfalls (as is urgent defecation, apparently), but add to those worries a pair of bleeding nipples, black and lost toenails, and becoming physically shorter due to lost fluid between your intervertebral disks, and the appeal really starts to wane.

So if you, like us, can't face the pain of 26.2 miles, but still desire the dopamine hit of the abundant 'likes' on a social media post, consider the Marathon du Médoc.

First held in 1985 in November, it's now run during a warmer September in France's Médoc wine region. The marathon hits the official 26.2 mark, but invites 'runners' on a route through vineyards, with roadside entertainment and exceptional food to indulge in along the way. Costumes are compulsory, as is the drinking, since par for the course is a sip of 23 famed vintages.

15 NOVEMBER

Birth of Aneurin Bevan (1897)

PENDERYN WELSH WHISKY –

After creating the British National Health Service, which provides healthcare for all, free at the point of delivery and based on need rather than wealth, Aneurin Bevan was entitled to a drink. And since he was Welsh, we suggest Penderyn.

The Penderyn Distillery is in the Brecon Beacons National Park, and their excellent wood maturation programme provides a wide range of varieties to dip into. The NHS is brilliant.

16 NOVEMBER

The Sound of Music Opens on Broadway (1959)

AUSTRIAN SCHNAPPS –

In the song 'My Favourite Things', Maria (played by Julie Andrews in the 1965 film) rhymes 'whiskers on kittens' with 'bright copper kettles and warm woollen mittens'. Mittens are a bit like gloves for thickos, but copper kettles are crucially employed in the production of schnapps in Austria.

Some schnapps can be sugary sweet and cheap, but Austria's crafted brandies are distilled in pot stills using only natural fruit, such as apples, apricots, lovely pears and big juicy plums.

17 NOVEMBER

PATENT GRANTED FOR FIRST COMPUTER MOUSE (1970)

MICKEY'S BIG MOUTH –

While embarking on a bid to create interactive computer systems to help humanity, Doug Engelbart introduced the world to word processing, document sharing and hyperlinks – and patented the first computer mouse.

For us, his crowning glory was the 'cut and paste' facility, which, combined with the invention of Wikipedia, has made this book possible.

Mickey's Big Mouth is a malt liquor brewed by Miller. Not the most discerning entry, but it was surprisingly tricky finding any mouse-related drinks.

18 NOVEMBER

TERRY WAITE RELEASED BY KIDNAPPERS (1986)

RAMOS GIN FIZZ –

Terry Waite was released on this day in 1986, having been held hostage for 1,736 days. And having been chained to a radiator for most of his captivity, the humanitarian might have discovered there's no better way to get blood rushing into the wrists than by making a Ramos Gin Fizz cocktail.

The drink was conceived by Henry Charles Ramos at the Imperial Cabinet Saloon in New Orleans in 1888, and became so popular in his next bar, The Stag, that he had 20 bartenders continuously creating them to meet demand. Requires rigorous and prolonged shaking.

~

60ml (2 fl oz) gin | 15ml (½ fl oz) lemon juice
15ml (½ fl oz) lime juice | 20ml (¾ fl oz) sugar syrup
25ml (¾ fl oz) single cream | 5ml (¼ fl oz) orange flower water
1 egg white | Ice | Soda water, to top up
Orange slice or zest, to garnish

- Shake the first seven ingredients hard in a cocktail shaker. Add the ice and shake hard again. Pour into a chilled highball glass, then pour a bit of soda water into the empty shaker, swirl to collect any leftovers and top the drink to boost the nice head. Garnish with an orange slice or zest.

---

19 NOVEMBER

LEWIS AND CLARK CAMP BY THE PACIFIC (1805)

JÄGERMEISTER –

Expeditionists Meriwether Lewis and William Clark mapped huge swathes of uncharted America, from Louisiana to the Pacific coast, between 1804 and 1806. And – get this – they undertook their epic journey with 130 barrels of whiskey in tow. Tied to carts at the expense of other vital supplies, they understood the spirit could bestow fortitude and morale.

Granted, they suffered everything from malaria and dysentery to rheumatism and frostbite, partly due to malnutrition and lack of suitable clothes, but they always had that whiskey.

Along with peaceful encounters with natives and discovering new species, their diarised accounts reveal they discovered a massive rock in Oregon and due to its phallic shape, christened it 'Cock Rock'. Because it looked like a huge penis.

By November 1805, the expedition had reached the Pacific and made camp by the sea, with diary entries on this day noting sea spray and a spot of deer hunting. So, while they probably paired American whiskey with freshly killed game meat, we suggest a slight detour in the drinks cabinet, towards Jägermeister.

The German word *jägermeister* translates to 'master hunter', and the liqueur, complete with 56 different herbs, roots, fruits and spices, was initially marketed in 1934 as a post-dinner tummy settler. It's actually a very complex creation, and best enjoyed chilled and neat.

---

20 NOVEMBER

FIRST MASS DEMONSTRATION DURING VELVET REVOLUTION (1989)

BLACK VELVET -

The Czech's Velvet or Gentle Revolution started after attacks on protesters by police on 17 November 1989. Rather than responding with violence, the entire nation turned the other cheek.

Shops closed, offices were emptied, flags were waved and polite chants were emitted. Thousands also shook keys every few minutes to mimic the bell that tolled for the Communist regime, and while that sound must've been hugely annoying, no one lost their temper.

By 29 November, the government was forced to end a one-party rule. Well done, everyone.

In their honour, enjoy a Black Velvet, first served at Brook's Club in London after Prince Albert's death in 1861, when staff suggested the blend represented champagne in mourning, dressed all in black.

~

100ml (3½ fl oz) Guinness | Champagne, to top up

- Pour the Guinness into a flute, then slowly top with champagne and gently stir.

21 NOVEMBER

ROCKY RELEASED (1976)

PHILADELPHIA BEER –

Costing $1 million to make, *Rocky* took $250 million at the box office, was an Oscar-winning success and inspired seven sequels. More importantly, there's a bar scene in the film.

While trying to sip his suds at the Lucky Seven Tavern, Rocky is pestered by a racist barkeep who questions the validity of the world champ Creed. Rocky becomes rightly irritated by the ignoramus, slams down his beer and storms out of the bar in anger. He drops his dollars for the beer as he leaves, because he's neither racist, nor thief.

Rocky's beer is a Schmidt's of Philadelphia, established in 1860 and, by the 1970s, the undefeated heavyweight of the city. When the brewery threw in the towel in 1987, there were no city-brewed challengers left, but from the gutter of defeat, the Philly beer scene legged it up the steps of success before a climactic and award-winning craft comeback.

Dock Street Brewery led that charge, now boasting a 35-year history and a selection of lager-style beers with flavour that should still satisfy the yellow fizzy Schmidt fans.

22 NOVEMBER

DEATH OF BLACK BEARD (1718)

STRONG RUM –

Infamous pirate Blackbeard was called Blackbeard *and* he had a black beard. It was a remarkable coincidence.

Along with ships, the swashbuckler sank a lot of rum – and some historians (us) have discovered he was the inventor of drinking games. His favourite involved getting shipmates drunk below deck before blowing out the candles and randomly firing his gun. There he was, in the dark, shooting salty seamen, and if he hit someone, that meant the crew member could not be trusted.

After only two years of high seas skulduggery, though, it was Blackbeard who came to a sticky end on this day in 1718. A surprise attack from the British navy off the coast of North Carolina had Blackbeard on the backfoot. He fought, but was shot five times and stabbed with more than 20 cutlasses.

After his death, his skull was turned into a punch bowl, which is pretty cool. So, if you have a skull to hand, enjoy some strong rum in it today.

---

23 NOVEMBER

SNOOP DOGG'S *DOGGYSTYLE* RELEASED (1993)

GIN AND JUICE –

As Snoop so wisely rapped, today is a day for 'rollin' down the street, smokin' indo, sippin' on gin and juice'. In the track 'Gin & Juice', his good friend Dr. Dre enters the fray with Tanqueray ('And a fat ass J'). So with this in mind, opt for London Dry Tanqueray, a quality classic gin. Launched in the 1830s, it is, much like Snoop, a pioneer in its field.

~

60ml (2 fl oz) Tanqueray gin | 50ml (1½ fl oz) orange juice
50ml (1½ fl oz) grapefruit juice | Ice

- Pour the gin and juices into a highball glass over ice. Stir, then garnish with chronic, hoes, that sort of carry-on.

24 NOVEMBER

'DISCOVERY' OF TASMANIA (1642)

TASMANIAN WHISKY –

Dutchman Abel Tasman became the first European to arrive in Tasmania. While he claimed it as a 'discovery', there were already some indigenous people there. The Europeans subsequently slaughtered nearly all of them. Arseholes.

While they also introduced alcohol, these same Europeans then banned distilling for more than 100 years. Honestly, could they have been any more arse-holey?

It wasn't until the 1990s, when Bill Lark convinced the government to see sense, that it was possible to obtaining a licence to distil again. Now his Lark Distillery, as well as the Old Hobart and Sullivan's Cove distilleries, are all producing whisky that is celebrated around the world.

25 NOVEMBER

SECOND PART OF *FEAR & LOATHING IN LAS VEGAS* SERIAL PUBLISHED IN *ROLLING STONE* (1971)

CHIVAS REGAL –

Hunter S. Thompson opens his classic novel *Fear & Loathing in Las Vegas* with a nifty list of narcotics, referencing everything from grass and mescaline to blotter acid, cocaine, uppers, downers, screamers and, indeed, laughers.

But fear not, because his protagonist didn't miss the alcohol. There was a quart of tequila and rum with a case of beer in the trunk, because Thompson loved a drink.

One recurring standard in the writer's life was Chivas Regal, the globally renown blended Scotch. The 12-year-old expression is a spicy, sweet and super-duper spirit, best sipped neat – it's too discerning to be mixed with Coke.

26 NOVEMBER

First Thanksgiving (1621)

PUMPKIN PATCH ALE, ROGUE –

In 1621, the Pilgrims were pleased as punch to discover the 'New World' had a plentiful supply of pumpkins, so they added them to the menu for the first Thanksgiving feed.

Pumpkins were not a 'discovery', of course. One of the oldest cultivated foods in human history, they had long been a crop for the indigenous residents, who the Europeans would go on to slaughter and steal pumpkin-arable land from.

Regardless, having made pies from pumpkins, European thoughts quickly turned to booze, thus fermenting the winter squash's juicy pulp into a rudimentary wine. By the 18th century, pumpkin beer was a standard in Europe, and courtesy of more recent craft brewer forays into traditional beer styles, there has been a resurgence.

Rogue use pumpkins grown at Rogue Farms in Independence, Oregon, and spice things up with orange peel, cinnamon, cloves, cardamom, vanilla, ginger and nutmeg.

27 NOVEMBER

Birth of Anders Celsius (1701)

FROZEN TOMMY'S MARGARITA –

An ideal cocktail temperature is -2°C, which is tricky to precisely measure without a thermometer. Anyone caught dipping temperature sensors into our drinks will get short shrift, but if you've had a kid in recent years, you might own a fancy digital device to clamp under an infant's armpit. Maybe repurpose it for cocktails.

We've got Anders Celsius to thank for all this measuring carry-on, he's the clever-clogs Swedish, white-coated boffin responsible for defining the international temperature scale.

If stirring a cocktail on ice in the US, then you might put more stock in Daniel Fahrenheit, and your ideal cocktail temp would be more like 28°F. Which, unlike a minus number, doesn't sound so cold really. So why don't you all just switch to °C?

~

60ml (2 fl oz) blanco tequila | 30ml (1 fl oz) lime juice
15ml (½ fl oz) agave syrup | Large handful of ice
Lime wedge, to garnish

- Blitz all ingredients (except the garnish) in a blender until smooth. Pour into a rocks glass and garnish with a lime wedge.

---

28 NOVEMBER

TRIAL OF ANNE BONNY (1720)

BAHAMA MAMA COCKTAIL –

Anne Bonny was one of the most fearsome female pirates in history, and made up a terrifying triumvirate with Calico Jack Rackham and Mary Read.

But things were far from plain sailing when their naughty nautical antics saw them captured in 1720. Setting off from Nassau, the Pirate Republic in the Bahamas, they were ship-faced on rum when surprised by pirate hunter Jonathan Barnet. Arresting them, he took the scurvy knaves to Jamaica, where they were quickly sentenced to swing from a rope.

Rackham was hung, but Bonny and Read secured a stay of execution after revealing they were pregnant. While Read died in prison, Bonny reportedly found freedom and lived on to deliver a bounty of eight more Bonny babies.

Like all the very best pirate tales, there is a pinch of sea salt sprinkled over Anne's escapades – by which we mean, of course, quite a lot of bullshit. Nonetheless, avast ye, because it's a fascinating addition to the annals of pirate tales, and it allows us to present a drink link more seamless than a main sail.

The Bahama Mama cocktail was created in 1961 by Oswald 'Slade' Greenslade, at the Nassau Beach Hotel.

~

15ml (½ fl oz) dark rum | 15ml (½ fl oz) overproof rum
15ml (½ fl oz) coconut liqueur | 15ml (½ fl oz) lemon juice
60ml (2 fl oz) pineapple juice | 7.5ml (¼ fl oz) coffee liqueur | Ice
Pineapple wedge, maraschino cherry and mint leaves, to garnish

- Shake all the ingredients (except garnishes) in a cocktail shaker with ice. Strain into an ice-filled tiki mug. Garnish with a pineapple wedge, maraschino cherry and mint leaves.

---

29 NOVEMBER

Release of *Pong* (1972)

BJ'S BREWHOUSE BLONDE –

There was a time when dweebs played video games without analog sticks. You couldn't button mash, or waggle, and the games had, like, no hope of asynchronous or asymmetric gameplay.

Crazy right? We have no idea what this means, but we do remember *Pong*.

While rudimentary in the face of the current console carnage, any committed gamers know Pong far from stinks – indeed, it deserves immeasurable respect. Brought to the screen by Allan Alcorn and Atari founder Nolan Bushnell, it was the first commercially successful video game and helped establish the industry.

Nolan Bushnell is also the creator of Chuck E. Cheese, family pizza restaurants with arcade machines, and that's a nice hybrid concept, but not as useful as BJ's, the southern Californian restaurant that combines pizzas with onsite microbrewing.

Amongst the best beers on offer is BJ's Brewhouse Blonde, a German *kölsch*, which is a fine companion to pizza thanks to a little malt and gentle hopping for bitter balance.

---

30 NOVEMBER

BIRTH OF WINSTON CHURCHILL (1874)

CARLSBERG SPECIAL BREW –

Whether you regard him as a national hero or a war-mongering wrong'un, it'd be daft not to have a drink in Winston Churchill's honour.

After all, it wasn't Churchill's war cabinet that won the Second World War: it was his copiously stocked drinks cabinet.

Churchill cherished champagne, his favourite being Pol Roger (who gave him his own personal stash in return for liberating France). He also kick-started every day with a glass of Johnnie Walker that would be topped up with water throughout the morning, and had an insatiable appetite for exquisite and exotic brandies, such as Hine Cognac and Armenian brandy sent to him by Stalin. The Danes even brewed a brandy-toned beer in his honour.

That beer is Carlsberg Special Brew. True story. So that's what we'll be having today. In a brandy snifter, while chomping on cigars and pushing little soldiers about a map with a windscreen wiper.

DECEMBER

1 DECEMBER

Franz Liszt's Debut (1822)

HUNGARIAN WINE –

In 1822, the 11-year-old pianist Franz Liszt debuted at the Landständischer Saal in Vienna. So celebrated was his performance that even Beethoven came over to say 'Wunderbar pianist' afterwards. 'Twas a big wow.

It's worth stressing he was 11 years old. Eleven. What were you doing when you were 11, eh? If you were a daydreaming adolescent boy, pianist might have been an option, but chances are you couldn't play one properly yet.

Liszt went on to become one of the greatest musicians of all time – and this despite some prolific drinking. He could drink a bottle of brandy a day and poured cognac on his watermelon, although this did account for some of his seedier behaviour, and inspired more melon-choly music. He also held great affection for his native Hungarian wine. Today, they produce lovely light Pinot Noirs and sweeter Tokaji.

2 DECEMBER

Death of Wojtek the Bear (1963)

TYSKIE –

Let us take paws to honour a beer-drinking bear who helped defeat the Nazis. The bear was named Wojtek, which means 'smiling warrior' who, having been abandoned by his fore-bears, was adopted as a cub by Polish soldiers travelling through Iran. The troops took him into their hearts, and he became a permanent member of the 22nd Artillery Supply Company.

Earning the rank of private, Wojtek boosted morale, learned to salute and even worked out how to shower. During the Battle of Monte Cassino in Italy, he carried live cannon ammo – bear-handed – while under fire; and when troops questioned spies, his appearance scared prisoners into spilling their guts. Such bravery earned him national hero status in Poland, and the 22nd Artillery Supply Company redesigned its logo to show Wojtek carrying an artillery shell.

Wojtek also loved a beer, and, after one or two, possibly served up in a massive growler, he was often found getting up to amusing japes: marching with the troops while saluting, and, on more than one occasion, stealing underwear from the washing lines of female soldiers. We've all done it. But no, come on, tsk tsk… Which brings us neatly onto Tyskie, the number-one selling Polish lager. Brewed in Tychy, a tiny town with an epic 400-year brewing heritage, Tyskie is an easy-drinking pale lager that would probably be endorsed by Polish bears.

---

3 DECEMBER

ELVIS'S COMEBACK SPECIAL (1968)

AKU AKU COCKTAIL –

According to *Rolling Stone*, Elvis's 1968 TV *Comeback Special* was watched by 42 per cent of the US viewing audience. The legend might have celebrated with a drink, although alcohol wasn't on his infamous list of overindulgences, with Elvis claiming he only once lost his way with alcohol, while drinking peach brandy. In his honour, then, a tropical nod to his love of Hawaii in the form of a twisted tiki special: Aku Aku.

~

15ml (½ fl oz) Briottet Crème de Pêche
45ml (1½ fl oz) fresh pineapple juice | 40ml (1¼ fl oz) sugar syrup
25ml (¾ fl oz) fresh lime juice | Ice
Mint leaves, pineapple slice and maraschino cherry, to garnish

- Blend all the ingredients (expect garnishes) in a blender on high speed. Pour into a cocktail glass and garnish with mint leaves, a pineapple slice and a maraschino cherry. Serve with a straw.

---

4 DECEMBER

MANILA PAPER PATENTED (1843)

PAPER PLANE –

In 1843, after seeing the writing on the wall for so long, John Mark and Lyman Hollingsworth of Delaware took the plunge and patented their Manila paper, which is still used for envelopes today.

Mercifully, they died before the advent of email, which might have left them wondering if paper is a bit shit. (Unless you're taking a shit, of course.) That said, paper has been used to make this book, so if you're reading it having purchased it full price from an actual bookshop, perhaps you should have a bit more respect.

Besides, paper is also useful for making paper planes, which still bring a sense of wonder to a toddler: so, if you're struggling with the book, repurpose the pages for planes. Mercifully, there's a drink called a Paper Plane to help this particularly tenuous entry hang together. The recipe was created by bartender Sam Ross.

~

25ml (¾ fl oz) bourbon | 25ml (¾ fl oz) amaro
25ml (¾ fl oz) Aperol | 25ml (¾ fl oz) lemon juice | Ice

- Shake all the ingredients in a cocktail shaker with ice, then strain into a coupe.

5 DECEMBER

Prohibition is repealed in the US (1933)

BATHTUB GIN –

Bars across America continue to celebrate Repeal Day, the end of Prohibition – and rightly so, since it was a seriously stupid idea.

For a start, Prohibition was almost impossible to enforce. Drinkers came up with ingenious ways to keep topped up: some imitated doctors or priests so they could get hold of medicinal alcohol or Communion wine; others hid illicit booze in everything from fake heels in their shoes to walking canes and hollowed-out eggs. No yolk.

Then there was the unregulated illegal bathtub moonshine, which was topped up with toxins and poisoned thousands.

You also had the organised crime epidemic, which was pretty bad (see 17 January).

6 DECEMBER

The Rolling Stones release 'Sympathy for the Devil' (1968)

DUVEL –

Back in 1968, this record sparked an infernal uproar, the biggest bone of contention being that lead singer Mick Jagger delivered his satanic spiel in the first person – that is to say, Sir Mick presented himself as the devil. Apparently, a few weirdos believed him.

So, here's an appropriate drink for the occasion: Duvel, Flemish for devil. The Belgian blonde beer boasts aromas of clove, pepper and spice, and delivers a gentle bitterness that makes it dangerously easy to drink.

The devil really is in the detail, though – specifically the bit that reads '8.5 per cent ABV'. Such strength seems unlikely as you sip the light and moreish beer, but that's how it acquired its reputation and name. So, please allow it to introduce itself to your mouth – but perhaps in moderation.

---

7 DECEMBER

JOHNNY WEISSMULLER BREAKS THE 150-YARD FREESTYLE WORLD RECORD (1925)

SCHNAPPS –

On this day in 1925, Johnny Weissmuller completed a 150-yard freestyle swim in 1 minute 25 seconds: one of an incredible 67 world records set by the five-time Olympic gold medallist. But he also played Tarzan, which is pretty cool. Added to which, he was firm friends with co-star Cheetah the chimpanzee.

There were various chimps used for the role, but perhaps the most notable Cheetah is the chap who sipped schnapps while finger-painting and listening to Christian music. That Cheetah lived to 80.

So, raise a glass to Johnny and Cheetah – and make it a fruit brandy. We recommend schnapps from St George, an American distillery set up in 1982 by Jörg Rupf, a craft distilling pioneer who uses traditional German techniques.

8 DECEMBER

Nuclear Arms Treaty Signed (1987)

VODKA MARTINIS –

Legend has it that in 1987, President Ronald Reagan and Mikhail Gorbachev signed the treaty banning the use of intermediate-range nuclear missiles, and then celebrated with vodka Martinis. The cocktail bridged the political divide between these two nations, combining the vodka of Russia with the cocktail of America. Did they do it, though? Did they really? We can't be sure – we weren't there.

~

60ml (2 fl oz) vodka | 5ml (¼ fl oz) vermouth | Ice
Lemon zest twist, to garnish

- Stir the vodka and vermouth in a mixing glass over ice, then strain into a chilled Martini glass. Garnish with a lemon twist.

9 DECEMBER

Birth of Fritz Maytag (1937)

ANCHOR STEAM BEER, ANCHOR BREWING –

In 1965, Fritz Maytag made an indelible glass stain on the global beer mat of craft brewing when he purchased a stake in San Francisco's Anchor Brewing, saving an ailing icon and inspiring a generation of beer geeks.

The Maytag family had form when it came to entrepreneurial success, having made a bundle and cleaned up in the washing machine industry, but Fritz got into a proper spin when he heard his favourite brewery was about to close.

In less than ten years he converted Anchor Brewing from dilapidated and desperate to daring and dynamic. He launched a porter in 1972, the only dark beer in America at the time, and became the first to create a seasonal brew in the form of Anchor Christmas Ale. Then, in 1975, he launched Liberty Ale, a hop-forward beer that was the forerunner to the now ubiquitous American IPA.

Anchor Steam remains the flagship beer, named after the mid-1800s San Francisco brewery rooftop fermenters that sent steam into the city's cold air. This smooth but citrusy session sipper is made according to the post-Prohibition recipe, using traditional shallow, open fermenters, and is a tribute to one of the few original American beer styles.

---

10 DECEMBER

DEATH OF ALFRED NOBEL (1896)

CARLSBERG –

Among Alfred Nobel's 330 patents was dynamite, but the invention really blew up in his face when a newspaper mistakenly ran his obituary, slagging him off for his deadly creation. Still very much alive when he read this obituary, Alfred's conscience was pricked, so he established the Nobel prize list in 1901.

The winners of the prize include Niels Bohr, who earned it in 1922 for revelations about atom structure. It was lab-coated boffinry we don't understand, but the important thing is Carlsberg gave the Dane a house as reward – and ran a pipe direct from the brewery to provide him with free beer until his death in 1962.

---

11 DECEMBER

WALL STREET RELEASED (1987)

EAST COAST AMERICAN SPARKLING WINE–

As Michael Douglas proved in the 1987 film *Wall Street*, city traders make a lot of 'cents' out of buying and selling, before they lob their lolly around, clapping their hands in the air and rudely shouting 'Champagne!' You could do that today, if you like, but when you do, demand some fizz from upstate New York. Finger Lakes County has several sparkling wine specialists, with the Dr Konstantin Frank Winery producing a celebrated rosé brut, and the Hermann J. Wiemer vineyard a cuvée brut with a traditional blend of 65 per cent Chardonnay and 35 per cent Pinot Noir.

12 DECEMBER

SATURDAY NIGHT FEVER RELEASED (1977)

SEVEN & SEVEN –

When John Travolta's Tony Manero struts into the Odyssey disco club in Brooklyn, New York, he grabs a table with his buddies and orders a Seven & Seven. This highball serve mixes blended American whiskey Seagrams 7 Crown with soft drink 7 Up. This helped catapult Seagrams 7 Crown to the number-one spirit brand in the 70s. Drink one and disco.

~

Ice | 30ml (1 fl oz) Seagrams 7 Crown
7 Up, or other lemon and lime soda
Lemon zest twist, to garnish

- Fill a highball glass with ice. Stir in the whiskey and top with 7 Up. Stir, then garnish with a lemon twist or some disco shapes.

13 DECEMBER

SIR FRANCIS DRAKE STARTS CIRCUMNAVIGATION OF GLOBE (1577)

MOJITO –

In 1577, Francis Drake became the first Englishman to circumnavigate the globe. No doubt there was a little rum and 'yo ho ho'-ing as he navigated the Caribbean, but rather than pulling corks out with teeth, glugging from bottles and calling each other aaaaarseholes, it transpires Drake's crew were a refined bunch who stirred up fancy drinks. Already identifying the medicinal benefits of Cuban mint and citrus, they reputedly blended them with local sugar-cane alcohol – so we can credit Drake with the creation of the Mojito. (While the historical foundation for this is as loose as the cocktail will make you feel, this is our book, so just go with it.)

~

60ml (2 fl oz) Cuban white rum | 15ml (½ fl oz) lime juice
10ml (½ fl oz) sugar syrup | 6 mint leaves
½ lime, chopped into chunks
Crushed ice | Soda water, to top up

- Pour the rum, lime juice and sugar syrup into a highball glass and add the mint leaves and lime chunks. Half-fill with ice and stir with a bar spoon. Add more ice, top with soda water and stir.

---

14 DECEMBER

DENNIS BERGKAMP MAKES HIS AJAX DEBUT (1986)

THE FLYING DUTCHMAN COCKTAIL –

Footballer Dennis Bergkamp had the sexshy shoccer shkills to pay for his outstanding utility bills, but while he occasionally flew down the wing, he flatly refused to board a plane.

Rather than being a pioneering eco-warrior/Greta Thunberg-type, though, Dennis was more like B. A. Baracus, in that his reason for not 'gettin' on no plane, fool' was a fear of flying. He subsequently earned nickname 'The Non-Flying Dutchman'.

An almost flawless link to a cocktail called the Flying Dutchman.

This creation comes from Tess Posthumus of the Flying Dutchmen Cocktails bar in Amsterdam.

~

45ml (1½ fl oz) Bols Barrel Aged Genever
30ml (1 fl oz) lemon juice | 30ml (1 fl oz) Monin Speculoos syrup
2 dashes of Regans' Orange Bitters
1 dash of The Bitter Truth Orange Flower Water | Ice
Orange zest twist and edible flower, to garnish

- Shake all the ingredients (except the garnishes) in a cocktail shaker with ice, then strain into a cocktail glass. Garnish with an orange zest twist and an edible flower.

---

15 DECEMBER

DEATH OF JERRY THOMAS (1885)

BLUE BLAZERR –

Known as the 'Professor', American bartender Jerry Thomas was author of the influential *Bar-Tender's Guide* and spent decades slinging drinks across the country. Such was his influence on 19th-century popular culture, his death made front-page news in New York.

The Blue Blazer was his signature flaming whisky drink. If you attempt to make it, pour the flaming alcohol between heatproof glassware – and keep it clear of your eyebrows, along with anything else you value.

~

Serves 2
120ml (4 fl oz) boiling water, plus extra to warm the mugs
120ml (4 fl oz) Scotch whisky
2 teaspoons demerara or raw sugar
Lemon zest twists, to garnish

- Preheat 2 glass mugs with some boiling water and gently heat the Scotch in a pan over a low heat.
- Pour the sugar, Scotch and measured boiling water into one of the

mugs, then light it. Pour the flaming liquid back and forth from mug to mug (about 5 times).
- Finally, pour into one mug and put out the flame. Garnish with lemon twists and serve.

16 DECEMBER

Birth of Madame Barbe-Nicole Clicquot (1777)

VEUVE CLICQUOT –

In 1805, a 27-year-old Madame Barbe-Nicole Clicquot inherited her dead husband's wine business. While it wasn't the done thing for a woman of this era to take a job, she grabbed the opportunity by the neck, shook up the Champagne industry and sprayed foam in the faces of her naysayers.

Now referred to as the 'Grande Dame of Champagne', Barbe-Nicole perfected her own brilliant bubbly, gave the world pink champagne, introduced a new sleek bottle shape, and created the very first Champagne vintage.

On top of which, she developed the riddling table, a pioneering device still used today that enables producers to collect and remove sediment from Champagne without impairing the quality, making Champagne clear and fresh.

17 DECEMBER

Birth of Angelo Mariani (1838)

VIN MARIANI –

Angelo Mariani was a 19th-century chemist who cleverly identified a method of extracting Peruvian coca leaf's punchy cocaine properties in a palatable form by mixing coca leaf, sugar, Bordeaux wine and brandy. Calling his drink Vin Tonique Mariani à la Coca de Pérou (catchy), the beverage became the easiest – and, indeed, tastiest – way to exploit the plant's narcotic qualities, delivering 21mg cocaine per 100ml (3½ fl oz) (fans of the drink included Thomas Edison – see 22 October). The tonic inspired Colonel John Pemberton to develop Coca-Cola, and, while the soft drink has since removed the cocaine, it delivered a dose right up until the early 20th century.

You'll find a modern Vin Mariani incarnation on sale today, wisely using top-growth Bordeaux fortified with de-cocainised Peruvian coca leaf.

18 DECEMBER

STANLEY MATTHEWS WINS THE BALLON D'OR (1956)

TITANIC STOUT –

Known as the 'Wizard of Dribble', footballer Stanley Matthews was the first ever winner of the now coveted Ballon d'Or. The extraordinary winger played until he was 50, won the FA Cup for Blackpool and was also the first footballer knighted. He started his career at Stoke City, so sip an award-winning stout from local brewery Titanic. Don't dribble it.

19 DECEMBER

A CHRISTMAS CAROL BY CHARLES DICKENS PUBLISHED (1843)

SMOKING BISHOP –

Drink infuses Dickens's stories, and the Smoking Bishop featured in *A Christmas Carol* is a quality and complex punch. While the name conjures a smug, post-coital man of the cloth, it was actually christened by revellers who poked fun at the church, sipping it while shirking Sunday duties.

The cocktail's luxury ingredients of wine and port made it a choice for the affluent, so the serving in Dickens's seasonal story also represents the pronounced Victorian class divide. As Scrooge offers his punch to the once poverty-stricken Bob Cratchit, Dickens uses the cocktail to hammer home his metaphor of how a reformed and wealthy protagonist should pour out kindness to the poor.

~

Serves 6

1 x 75cl bottle Ruby port | 1 x 75cl bottle red wine
6 oranges | 2 lemons | 30 cloves
½ teaspoon ground allspice | ½ teaspoon ground cinnamon
5cm (2in) piece fresh root ginger, chopped
65g (2¼ oz) demerara sugar | 5 dried cardamom pods

- Preheat the oven to 180°C/gas mark 4/350°F. Stud the oranges and a lemon with 5 cloves in each and roast for an hour. Place all the other ingredients in pan on a low heat. While they simmer, cut and squeeze the cooked fruit and add the juice to the pan. When warmed through, strain into a punch bowl and ladle into individual punch cups. Garnish with orange zest twists.

---

20 DECEMBER

IT'S A WONDERFUL LIFE RELEASED (1946)

HUDSON BABY BOURBON,
TUTHILLTOWN SPIRITS –

During one scene of Frank Capra's 1946 seasonal tear-inducer, James Stewart's character George Bailey visits Nick's Bar and orders a double bourbon, which, frankly, is all we need for a link here. Since he's in Upstate New York, we suggest Tuthilltown's Hudson Baby Bourbon. Set up in 2003, Tuthilltown created the first legally distilled and aged New York grain spirit since Prohibition.

21 DECEMBER

WORLD MONKEY DAY

MONKEY GLAND COCKTAIL –

In the early 20th century, Serge Voronoff, French surgeon and Nobel-Prize recipient, no less, was exploring methods for halting the ageing process. He was also immersed in animal-to-human transplants, believing the transfer of animal hormones could cure certain human ailments. Naturally, his studies soon turned to the monkey, because the monkey is the funniest and best of all the animals.

Serge held the specific belief that a monkey testicle was the key to youth, and convinced the medical community – and rather a lot of patients – that testosterone was vital for a healthy life. Thus, he grafted monkey testicles onto elderly Frenchmen. Actual monkey gonads, onto actual men.

Despite the strange nature of this treatment (and the fact that he somehow dropped the other ball on a proposed horse penis transplant), Serge subsequently made rather a lot of money out of the technique. Sadly, the procedure was later ridiculed, because, well, it doesn't work. Trust us.

Rather ironically, Serge became the butt of jokes, but his work might have coined negative medical terminology including 'that's nuts', 'balls to that' and, indeed, 'bollocks'.

So, in honour of his bizarre procedure, a witty bartender by the name of Harry MacElhone came up with the Monkey Gland cocktail in his Harry's New York Bar in Paris. MacElhone was no slouch when it came to slinging drinks, so fear not: the Monkey Gland is a tasty drink and *doesn't* taste like ball sacks.

~

60ml (2 fl oz) dry gin | 45ml (1½ fl oz) orange juice
5ml (¼ fl oz) absinthe | 5ml (¼ fl oz) grenadine
5ml (¼ fl oz) sugar syrup (optional) | Canned lychee, to garnish

- Shake all the ingredients (except the garnish) in a shaker with ice, then strain into a chilled cocktail glass or coupe. For a garnish, Harry suggested a slice of orange peel, but we think a tinned lychee is more appropriate. And serve it next to monkey nuts.

---

22 DECEMBER

FYODOR DOSTOEVSKY AVOIDS EXECUTION (1849)

STOLICHNAYA VODKA –

Dostoevsky's literary antics led to him being accused of anti-government sentiment in 1849, for which he was sentenced to death. At the last moment, however, he enjoyed a reprieve and was sent to a labour camp instead. He struggled to celebrate with a drink in the clink, but once freed, the writer sipped cognac before dessert and homegrown grain vodka before breakfast. Granted, he was a great literary mind, but there's nothing big or clever about vodka for breakfast.

---

23 DECEMBER

BIRTH OF DONNA TARTT (1963)

THE GOLDFINCH –

Happy birthday Donna Tartt, the exceptional author of books including *The Secret History* and Pulitzer prize-winner *The Goldfinch.* We love her work and are confident this will mean the world to her as she seeks out this entry in ours. She's reading this, we all know she is.

Lauren Shell's cocktail shares the name with Donna's book and is served at Seaworthy bar in New Orleans.

~

30ml (1 fl oz) Cocchi Americano | 30ml (1 fl oz) fino sherry
20ml (¾ fl oz) fresh lemon juice | 15ml (½ fl oz) sugar syrup
2 dashes orange bitters | Ice | Sparkling water
Grapefruit zest twist, to garnish

- Shake the first 5 ingredients in a cocktail shaker with ice, then strain into a chilled Collins glass filled with ice and top with sparkling water. Garnish with a grapefruit peel.

---

24 DECEMBER

Christmas Eve

TURKISH WINE –

More than 2 billion people celebrate Christmas, and many leave a drink out for the fat red berk who breaks into their gaff, with traditional treats ranging from sherry in the UK, Guinness in Ireland, cold lager in Australia and *akvavit* in Sweden.

But he probably doesn't even *like* these drinks – there's no reason he would. He's not even real. Besides, Saint Nicholas, the original Santa Claus, was actually a Greek bishop who preached in Turkey.

So, if you insist, why not leave him Turkish wine tonight? Turkey is the world's fourth-biggest grape producer, with 1,500,000 acres of vineyards, and the Romans were lauding its terroir 4,000 years ago.

---

25 DECEMBER

Christmas Day

BEER –

Pour some beer instead of wine with your Christmas din-dins this year. Why? Because it's what the baby Jesus would've wanted. Don't believe us? Well, for proof the Messiah was a beer guy, simply look at the location of ancient Israel, where he lived. It was flanked on either side by Egypt and Mesopotamia, both of which were hotbeds of beer and brewing.

Added to which, references to beer and beer-drinking are rife in the Hebrew Bible. Yahweh, God of Israel and the Judah kingdoms, drinks around two litres of beer every day, and moderate beer drinking is recommended throughout (see Isaiah 5:11, 28:7 and Proverbs 20:1, 31:4).

We associate the big man with wine thanks to an etymological bone of contention centring on the Hebrew word '*shekhar*', meaning 'strong drink'. Many Bible students assumed it to be a reference to wine, but *shekhar* relates to the word *sikaru*, an ancient Semitic term meaning 'barley beer'.

Now, we accept there are brainiacs (actual learned characters with glasses and that) who will contest these language issues. But let's be honest: the real reason beer was banished from subsequent versions of the Bible is sheer scholarly snobbery. When the Bible was translated into English in the early 17th century, beer was considered the drink of the pauper, while wine was popular among posh folk. So, in an astonishing display of academic arrogance, translators took it upon themselves to transform Jesus Christ from beer-drinking friend of the people into a posturing nouveau-riche wine-drinking playboy.

We all know that's not how Jesus would have rolled. You only have to look at him: he had a beard, he wore sandals, and he used to hang around with other men who had beards and wore sandals. If you've been to a real ale festival, you'll know this is the clobber they prefer.

And if you still don't believe us, where do you think we get the word '*He-Brew*'?

26 DECEMBER

BOXING DAY FOOTBALL

LUKAS HELLES, THORNBRIDGE –

Boxing Day has been a firm football fixture date in the UK for more than 150 years, but the first to sacrifice their Christmas Day excesses for our entertainment were Sheffield sides Hallam F.C. and Sheffield F.C. in 1860.

Today, the city is a breeding ground for some of the best British craft brewing, the elder statesmen of the new wave being Thornbridge. As well as excellent beer, the brewery also owns the Coach & Horses pub opposite Sheffield F.C., the oldest football club in the world.

So we suggest their Lukas Helles. Traditionally lagered and brewed entirely with Bavarian ingredients, it's a worthy substitute for any pre-match cooking lager.

27 DECEMBER

LOUIS PASTEUR BORN (1822)

WILD BEER –

When Louis Pasteur peered into his lunchtime pint and realised it was turning sour, he didn't simply ask for another pint, he was a nerd after all. Instead, he managed to deduce how the spoiling might be curtailed if the drink's temperature was raised for a certain amount of time, thus destroying the pathogenic microorganisms, or bacteria, upsetting his palate.

So it was his pint that led science on a path of pasteurisation across the nation, meaning he used beer to save millions of lives.

28 DECEMBER

Birth of Stan Lee (1922)

SPIDER MONKEY ORGANIC UNFILTERED IPA –

Listen up, webslingers! In the *Web of Spider-Man* issue #38, Peter Parker unwittingly sinks seven glasses of a spiked fruit punch at a party before drunk-fighting his nemesis the Hobgoblin. This proper post-pub boozy scrap sees windows smashed, bins knocked over, kebabs spilled and people flying on Goblin gliders. Our hero survives, but only just, proving even superheroes need to drink less but better.

Black Isle Brewing Co.'s Spider Monkey is an excellent beer, has the word 'spider' in there, and is a juicy IPA boasting more tropical fruit flavours than a badly made spiked punch.

29 DECEMBER

Tycho Brahe Fights a Duel (1566)

FRUITS OF THE SUN, MIKKELLER –

Tycho Brahe was a 16th-century Danish astronomer whose studies revealed how the planets revolved around the sun.

He came from Aarhuus (pronounced 'aarhoose') on the east coast, and lived with a moose who drank Danish lager before running around his mansion causing havoc. From this, we can deduce that, in the mid-1500s, it is likely there would have been a moose loose aboot Aarhuus.

Brahe also wore a metal nose after losing his real one in a drunken duel with a fellow astronomer on this day in 1566. How did he smell? F*cking awful.

Fruits of the Sun is brewed by the Danish craft stars Mikkeller. It has the word 'sun' in its name, and is a sour, fruited Berliner Weisse, which, assuming you don't have a metal nose, smells and tastes like wine gums.

30 DECEMBER

USSR FORMED (1922)

BELUGA VODKA -

In 1922, a brand-new communist state based on Marxist socialism emerged from the East – and the rest, as they say, is history. It's a history that fills up most of the 20th century, in fact, so don't expect us to succinctly sum it up here. Instead, pour yourself a measure of Beluga vodka, a luxury spirit that has sponsored the Russian Polo team, making it far removed from the objectives of the now-defunct USSR.

31 DECEMBER

NEW YEAR'S EVE

Here's a Greenwich Mean Time sipping schedule to help you see in midnight with the rest of the planet.

Midday: As midnight spreads across the Southern Hemisphere, work through New Zealand's Yeastie Boys Gunnamatta, an IPA with Earl Grey tea, for a first brew of the day. Or, for Australia, opt for Four Pillars Bloody Shiraz, gin steeped with Victorian shiraz grapes.

3–4pm: East Asia. For Japan, raise some sake, for flip's sake; and for China, try their Snow beer, the best-selling beer in the world.

6pm: As New Year reaches South Asia, pour mead for India. Sanskrit texts reference post-wedding rituals of honey beer, giving us the term 'honeymoon'.

8pm: They don't drink much in the Middle East, so take a break and have a milk. Sheik.

11pm: For central Europe, celebrate with vermouth, a fortified wine made by the French, Italians and Germans, and flavoured with botanicals from all over the continent.

Midnight: UK and Casablanca. Raise a gin Martini to the gin joints of Casablanca.

1am: Greenland. The ancient Arctic Inuits gulped gull wine made from a dead bird left in water to ferment. Not a drink worth getting in a flap about.

3am: Rio. Carnival with a caipirinha.

4am: The Americas. For Chile, sip a meaty Malbec (although this is a bold choice as you start to seriously consider bedtime). For New York and Havana, have a rum Manhattan.

6am: Mexico City. Mezcal. You might as well.

6.05am: Go to bed.

Noon: Midnight falls on the Pacific on Baker Island, an uninhabited US atoll that seasonally hosts the weary wings of ruddy turnstones and bar-tailed godwits, who will be very sober. As should you be. Now, get on with the New Year in a more responsible fashion. Drink less, but drink better. And remember, there's no sense in giving drink up in January – it's by far the grimmest month of the year.

ACKNOWLEDGEMENTS

Tom

Thanks to my beautiful and patient wife Claire for supporting me as I drink (less but better) for a living. The relentlessly inquisitive Joseph and Samuel, for love, cuddles and showing interest in monkey 'peanuts' being sewed onto the face of a French man. Mum & Dad for love, support and creating my brain. For all your ongoing support Ellen and Edward, Nikki, Ida and Margot; along with Rob, Janet, Stuart, Tracy, Harry & Rory: may you all buy this book. To my friends, I do have some. Ben McFartland, the wind beneath my wings. And a big 'well done' to the Internet.

Ben

Thanks to my red-hot, smokin' wife Sophie – you're my first choice; and our gorgeous, beautiful boys Rémy & Rory whose editorial 'assistance' helped prolong the book writing 'journey'. A special thanks to my lovely Mum for her endless love, support and proofing skills. Thanks to Dad and Nicola for their love and encouragement; big up to big brother Barnaby & Rosie Green, Baba, the Le Bars crew, the Gedye gang, NLC, the Dallat massive and, of course, Tom Sandham – if this book does well, we can write another one. Skills.

Colossal gratitude to everyone at Kyle Books including the wonderful Joanna Copestick – for backing us yet again, the incredibly patient Louise McKeever (apologies for the slight over-write); Tara O'Sullivan for spotting our numerous errors and indulgences, the legal team – we love you guys; designer Paul Palmer-Edwards, Pete Hunt in production and both Megan Brown and Charlotte Saunders for bringing the noise. Thanks also to Ben Clark of The Soho Agency and our agent Andy Townsend for helping make this happen.